Evolution in Modern Thought

by Ernst Haeckel and J. Arthur Thomson and August Weismann

EVOLUTION IN MODERN THOUGHT

I

DARWIN'S PREDECESSORS

BY J. ARTHUR THOMSON

Professor of Natural History in the University of Aberdeen

In seeking to discover Darwin's relation to his predecessors it is useful to distinguish the various services which he rendered to the theory of organic evolution.

(I) As everyone knows, the general idea of the Doctrine of Descent is that the plants and animals of the present day are the lineal descendants of ancestors on the whole somewhat simpler, that these again are descended from yet simpler forms, and so on backwards towards the literal "Protozoa" and "Protophyta" about which we unfortunately know nothing. Now no one supposes that Darwin originated this idea, which in rudiment at least is as old as Aristotle. What Darwin did was to make it current intellectual coin. He gave it a form that commended itself to the scientific and public intelligence of the day, and he won widespread conviction by showing with consummate skill that it was an effective formula to work with, a key which no lock refused. In a scholarly, critical, and pre-eminently fair-minded way, admitting difficulties and removing them, foreseeing objections and forestalling them, he showed that the doctrine of descent supplied a modal interpretation of how our present-day fauna and flora have come to be.

(II) In the second place, Darwin applied the evolution-idea to particular problems, such as the descent of man, and showed what a powerful organon it is, introducing order into masses of uncorrelated facts, interpreting enigmas both of structure and function, both bodily and mental, and, best of all, stimulating and guiding further investigation. But here again it cannot be claimed that Darwin was original. The problem of the descent or ascent of man, and other particular cases of evolution, had attracted not a few naturalists before Darwin's day, though no one [except Herbert Spencer in the psychological domain (1855)] had come near him in precision and thoroughness of inquiry.

(III) In the third place, Darwin contributed largely to a knowledge of the factors in the evolution-process, especially by his analysis of what occurs in the case of domestic animals and cultivated plants, and by his elaboration of the theory of Natural Selection which Alfred Russel Wallace independently stated at the same time, and of which there had been a few previous suggestions of a more or less vague description. It was here that Darwin's originality was greatest, for he revealed to naturalists the many different forms--often very subtle--which natural selection takes, and with the insight of a disciplined scientific imagination he realised what a mighty engine of progress it has been and is.

(IV) As an epoch-marking contribution to Natural History in the widest sense, we rank the picture which Darwin gave to the world of the web of life, that is to say, of the inter-relations and linkages in Nature. For the Biology of the individual--if that be not a contradiction in terms--no idea is more fundamental than that of the correlation of organs, but Darwin's most characteristic contribution was not less fundamental,--it was the idea of the correlation of organisms. This, again, was not novel; we find it in the works of naturalists like Christian Conrad Sprengel, Gilbert White, and Alexander von Humboldt, but the realisation of its full import was distinctly Darwinian.

As Regards the General Idea of Organic Evolution

While it is true, as Prof. H. F. Osborn puts it, that "'Before and after Darwin' will always be the ante et post urbem conditam of biological history," it is also true that the general idea of organic evolution is very ancient. In his admirable sketch From the Greeks to Darwin,[1] Prof. Osborn has shown that several of the ancient philosophers looked upon Nature as a gradual development and as still in process of change. In the suggestions of Empedocles, to take the best instance, there were "four sparks of truth,--first, that the development of life was a gradual process; second, that plants were evolved before animals; third, that imperfect forms were gradually replaced (not succeeded) by perfect forms; fourth, that the natural cause of the production of perfect forms was the extinction of the imperfect."[2] But the fundamental idea of one stage giving origin to another was absent. As the blue Aegean teemed with treasures of beauty and threw many upon its shores, so did Nature produce like a fertile artist what had to be rejected as well as what was able to survive, but the idea of one species emerging out of another was not yet conceived.

Aristotle's views of Nature[3] seem to have been more definitely evolutionist than those of his predecessors, in this sense, at least, that he recognised not only an ascending scale, but a genetic series from polyp to man and an age-long movement towards perfection. "It is due to the resistance of matter to form that Nature can only rise by degrees from lower to higher types." "Nature produces those things which, being continually moved by a certain principle contained in themselves, arrive at a certain end."

To discern the outcrop of evolution-doctrine in the long interval between Aristotle and Bacon seems to be very difficult, and some of the instances that have been cited strike one as forced. Epicurus and Lucretius, often called poets of evolution, both pictured animals as arising directly out of the earth, very much as Milton's lion long afterwards pawed its way out. Even when we come to Bruno who wrote that "to the sound of the harp of the Universal Apollo (the World Spirit), the lower organisms are called by stages to higher, and the lower stages are connected by intermediate forms with the higher," there is great room, as Prof. Osborn points out,[4] for difference of opinion as to how far he was an evolutionist in our sense of the term.

The awakening of natural science in the sixteenth century brought the possibility of a concrete evolution theory nearer, and in the early seventeenth century we find evidences of a new spirit--in the embryology of Harvey and the classifications of Ray. Besides sober naturalists there were speculative dreamers in the sixteenth and seventeenth centuries who had at least got beyond static formulae, but, as Professor Osborn points out,[5] "it is a very striking fact, that the basis of our modern methods of studying the Evolution problem was established not by the early naturalists nor by the speculative writers, but by the Philosophers." He refers to Bacon, Descartes, Leibnitz, Hume, Kant, Lessing, Herder, and Schelling. "They alone were upon the main track of modern thought. It is evident that they were groping in the dark for a working theory of the Evolution of life, and it is remarkable that they clearly perceived from the outset that the point to which observation should be directed was not the past but the present mutability of species, and further, that this mutability was simply the variation of individuals on an extended scale."

Bacon seems to have been one of the first to think definitely about the mutability of species, and he was far ahead of his age in his suggestion of what we now call a Station of Experimental Evolution. Leibnitz discusses in so many words how the species of animals may be changed and how intermediate species may once have linked those that now seem discontinuous. "All natural orders of beings present but a single chain".... "All advances by degrees in Nature, and nothing by leaps." Similar evolutionist statements are to be found in the works of the other "philosophers," to whom Prof. Osborn refers, who were, indeed, more scientific than the naturalists of their day. It must be borne in mind that the general idea of organic evolution--that the present is the child of the past--is in great part just the idea of human history projected upon the natural world, differentiated by the qualification that the continuous "Becoming" has been wrought out by forces inherent in the organisms themselves and in their environment.

A reference to Kant[6] should come in historical order after Buffon, with whose writings he was acquainted, but he seems, along with Herder and Schelling, to be best regarded as the culmination of the evolutionist philosophers--of those at least who interested themselves in scientific problems. In a famous passage he speaks of "the agreement of so many kinds of animals in a certain common plan of structure" ... an "analogy of forms" which "strengthens the supposition that they have an actual blood-relationship, due to derivation from a common parent." He speaks of "the great Family of creatures, for as a Family we must conceive it, if the above-mentioned continuous and connected relationship has a real foundation." Prof. Osborn alludes to the scientific caution which led Kant, biology being what it was, to refuse to entertain the hope "that a Newton may one day arise even to make the production of a blade of grass comprehensible, according to natural laws ordained by no intention." As Prof. Haeckel finely observes, Darwin rose up as Kant's Newton.[7]

The scientific renaissance brought a wealth of fresh impressions and some freedom from the tyranny of tradition, and the twofold stimulus stirred the speculative activity of a great variety of men from old Claude Duret of

Moulins, of whose weird transformism (1609) Dr. Henry de Varigny[8] gives us a glimpse, to Lorenz Oken (1779-1851) whose writings are such mixtures of sense and nonsense that some regard him as a far-seeing prophet and others as a fatuous follower of intellectual will-o'-the-wisps. Similarly, for De Maillet, Maupertuis, Diderot, Bonnet, and others, we must agree with Professor Osborn that they were not actually in the main Evolution movement. Some have been included in the roll of honour on very slender evidence, Robinet for instance, whose evolutionism seems to us extremely dubious.[9]

The first naturalist to give a broad and concrete expression to the evolutionist doctrine of descent was Buffon (1707-1788), but it is interesting to recall the fact that his contemporary Linneus (1707-1778), protagonist of the counter-doctrine of the fixity of species,[10] went the length of admitting (in 1762) that new species might arise by inter-crossing. Buffon's position among the pioneers of the evolution-doctrine is weakened by his habit of vacillating between his own conclusions and the orthodoxy of the Sorbonne, but there is no doubt that he had firm grasp of the general idea.

Erasmus Darwin (1731-1802), probably influenced by Buffon, was another firm evolutionist, and the outline of his argument in the Zoonomia[11] might serve in part at least to-day. "When we revolve in our minds the metamorphoses of animals, as from the tadpole to the frog; secondly, the changes produced by artificial cultivation, as in the breeds of horses, dogs, and sheep; thirdly, the changes produced by conditions of climate and of season, as in the sheep of warm climates being covered with hair instead of wool, and the hares and partridges of northern climates becoming white in winter: when, further, we observe the changes of structure produced by habit, as seen especially in men of different occupations; or the changes produced by artificial mutilation and prenatal influences, as in the crossing of species and production of monsters; fourth, when we observe the essential unity of plan in all warm-blooded animals,--we are led to conclude that they have been alike produced from a similar living filament".... "From thus meditating upon the minute portion of time in which many of the above changes have been produced, would it be too bold to imagine, in the great length of time since the earth began to exist, perhaps millions of years before the commencement of the history of mankind, that all warm-blooded animals have arisen from one living filament?"... "This idea of the gradual generation of all things seems to have been as familiar to the ancient philosophers as to the modern ones, and to have given rise to the beautiful hieroglyphic figure that is, by Divine Love; from whence proceeded all things which exist."

Lamarck (1744-1829) seems to have become an evolutionist independently of Erasmus Darwin's influence, though the parallelism between them is striking. He probably owed something to Buffon, but he developed his theory along a different line. Whatever view be held in regard to that theory there is no doubt that Lamarck was a thorough-going evolutionist. Professor Haeckel speaks of the Philosophie Zoologique as "the first connected and thoroughly logical exposition of the theory of descent."[12]

Besides the three old masters, as we may call them, Buffon, Erasmus Darwin, and Lamarck, there were other quite convinced pre-Darwinian evolutionists. The historian of the theory of descent must take account of Treviranus whose Biology or Philosophy of Animate Nature is full of evolutionary suggestions; of Etienne Geoffroy St. Hilaire, who in 1830, before the French Academy of Sciences, fought with Cuvier, the fellow-worker of his youth, an intellectual duel on the question of descent; of Goethe, one of the founders of morphology and the greatest poet of Evolution--who, in his eighty-first year, heard the tidings of Geoffrey St. Hilaire's defeat with an interest which transcended the political anxieties of the time; and of many others who had gained with more or less confidence and clearness a new outlook on Nature. It will be remembered that Darwin refers to thirty-four more or less evolutionist authors in his Historical Sketch, and the list might be added to. Especially when we come near to 1858 do the numbers increase, and one of the most remarkable, as also most independent champions of the evolution-idea before that date was Herbert Spencer, who not only marshalled the arguments in a very forcible way in 1852, but applied the formula in detail in his Principles of Psychology in 1855.[13]

It is right and proper that we should shake ourselves free from all creationist appreciations of Darwin, and that we should recognise the services of pre-Darwinian evolutionists who helped to make the time ripe, yet one cannot help feeling that the citation of them is apt to suggest two fallacies. It may suggest that Darwin simply entered into the labours of his predecessors, whereas, as a matter of fact, he knew very little about them till after he had been for years at work. To write, as Samuel Butler did, "Buffon planted, Erasmus Darwin and Lamarck watered, but it was Mr. Darwin who said 'That fruit is ripe,' and shook it into his lap" ... seems to us a quite misleading version of the facts of the case. The second fallacy which the historical citation is a little apt to suggest is that the filiation of ideas is a simple problem. On the contrary, the history of an idea, like the pedigree of an organism, is often very intricate, and the evolution of the evolution-idea is bound up with the whole progress of the world. Thus in order to interpret Darwin's clear formulation of the idea of organic evolution and his convincing presentation of it, we have to do more than go back to his immediate predecessors, such as Buffon, Erasmus Darwin, and Lamarck; we have to inquire into the acceptance of evolutionary conceptions in regard to other orders of facts, such as the earth and the solar system;[14] we have to realise how the growing success of scientific interpretation along other lines gave confidence to those who refused to admit that there was any domain from which science could be excluded as a trespasser; we have to take account of the development of philosophical thought, and even of theological and religious movements; we should also, if we are wise enough, consider social changes. In short, we must abandon the idea that we can understand the history of any science as such, without reference to contemporary evolution in other departments of activity.

While there were many evolutionists before Darwin, few of them were expert naturalists and few were known outside a small circle; what was of much more importance was that the genetic view of Nature was insinuating itself in regard to other than biological orders of facts, here a little and there a little, and that the scientific spirit had ripened since the days when Cuvier laughed Lamarck out of court. How was it that Darwin succeeded where others had failed? Because, in the first place, he had clear visions which a University curriculum had not made impossible, which the Beagle voyage made vivid, which an unrivalled British doggedness made real--visions of the web of life, of the fountain of change within the organism, of the struggle for existence and its winnowing, and of the spreading genealogical tree. Because, in the second place, he put so much grit into the verification of his visions, putting them to the proof in an argument which is of its kind--direct demonstration being out of the question--quite unequalled. Because, in the third place, he broke down the opposition which the most scientific had felt to the seductive modal formula of evolution by bringing forward a more plausible theory of the process than had been previously suggested. Nor can one forget, since questions of this magnitude are human and not merely academic, that he wrote so that all men could understand.

As Regards the Factors of Evolution

It is admitted by all who are acquainted with the history of biology that the general idea of organic evolution as expressed in the Doctrine of Descent was quite familiar to Darwin's grandfather and to others before and after him, as we have briefly indicated. It must also be admitted that some of these pioneers of evolutionism did more than apply the evolution-idea as a modal formula of becoming, they began to inquire into the factors in the process. Thus there were pre-Darwinian theories of evolution, and to these we must now briefly refer.[15]

In all biological thinking we have to work with the categories Organism--Function--Environment, and theories of evolution may be classified in relation to these. To some it has always seemed that the fundamental fact is the living organism,--a creative agent, a striving will, a changeful Proteus, selecting its environment, adjusting itself to it, self-differentiating and self-adaptive. The necessity of recognising the importance of the organism is admitted by all Darwinians who start with inborn variations, but it is open to question whether the whole truth of what we might call the Goethian position is exhausted in the postulate of inherent variability.

To others it has always seemed that the emphasis should be laid on Function,--on use and disuse, on doing and not doing. Practice makes perfect; c'est ?force de forger qu'on devient forgeron. This is one of the fundamental ideas of Lamarckism; to some extent it met with Darwin's approval; and it finds many supporters to-day. One of the ablest of these--Mr. Francis Darwin--has recently given strong reasons for combining a modernised Lamarckism with what we usually regard as sound Darwinism.[16]

To others it has always seemed that the emphasis should be laid on the Environment, which wakes the organism to action, prompts it to change, makes dints upon it, moulds it, prunes it, and finally, perhaps, kills it. It is again impossible to doubt that there is truth in this view, for even if environmentally induced "modifications" be not transmissible, environmentally induced "variations" are; and even if the direct influence of the environment be less important than many enthusiastic supporters of this view--may we call them Buffonians--think, there remains the indirect influence which Darwinians in part rely on,--the eliminative process. Even if the extreme view be held that the only form of discriminate elimination that counts is inter-organismal competition, this might be included under the rubric of the animate environment.

In many passages Buffon[17] definitely suggested that environmental influences--especially of climate and food--were directly productive of changes in organisms, but he did not discuss the question of the transmissibility of the modifications so induced, and it is difficult to gather from his inconsistent writings what extent of transformation he really believed in. Prof. Osborn says of Buffon: "The struggle for existence, the elimination of the least-perfected species, the contest between the fecundity of certain species and their constant destruction, are all clearly expressed in various passages."

Erasmus Darwin[19] had a firm grip of the "idea of the gradual formation and improvement of the Animal world," and he had his theory of the process. No sentence is more characteristic than this: "All animals undergo transformations which are in part produced by their own exertions, in response to pleasures and pains, and many of these acquired forms or propensities are transmitted to their posterity." This is Lamarckism before Lamarck, as his grandson pointed out. His central idea is that wants stimulate efforts and that these result in improvements which subsequent generations make better still. He realised something of the struggle for existence and even pointed out that this advantageously checks the rapid multiplication. "As Dr. Krause points out, Darwin just misses the connection between this struggle and the Survival of the Fittest."[20]

Lamarck[21] (1744-1829) seems to have thought out his theory of evolution without any knowledge of Erasmus Darwin's which it closely resembled. The central idea of his theory was the cumulative inheritance of functional modifications. "Changes in environment bring about changes in the habits of animals. Changes in their wants necessarily bring about parallel changes in their habits. If new wants become constant or very lasting, they form new habits, the new habits involve the use of new parts, or a different use of old parts, which results finally in the production of new organs and the modification of old ones." He differed from Buffon in not attaching importance, as far as animals are concerned, to the direct influence of the environment, "for environment can effect no direct change whatever upon the organisation of animals," but in regard to plants he agreed with Buffon that external conditions directly moulded them.

Treviranus[22] (1776-1837), whom Huxley ranked beside Lamarck, was on the whole Buffonian, attaching chief importance to the influence of a changeful environment both in modifying and in eliminating, but he was also Goethian, for instance in his idea that species like individuals pass through periods of growth, full bloom, and decline. "Thus, it is not only the great catastrophes of Nature which have caused extinction, but the completion of cycles of existence, out of which new cycles have begun." A characteristic sentence is quoted by Prof. Osborn: "In every living being there exists a capability of an endless variety of form-assumption; each possesses the power to adapt its organisation to the changes of the outer world, and it is this power, put into action by the change of the universe, that has raised the simple zoophytes of the primitive world to continually higher stages of organisation, and has introduced a countless variety of species into animate Nature."

Goethe[23] (1749-1832), who knew Buffon's work but not Lamarck's, is peculiarly interesting as one of the first to use the evolution-idea as a guiding hypothesis, e.g. in the interpretation of vestigial structures in man, and to realise that organisms express an attempt to make a compromise between specific inertia and individual change. He gave the finest expression that science has yet known--if it has known it--of the kernel-idea of what is called "bathmism," the idea of an "inherent growth-force"--and at the same time he held that "the way of life powerfully reacts upon all form" and that the orderly growth of form "yields to change from externally acting causes."

Besides Buffon, Erasmus Darwin, Lamarck, Treviranus, and Goethe, there were other "pioneers of evolution," whose views have been often discussed and appraised. Geoffroy Saint-Hilaire (1772-1884), whose work Goethe so much admired, was on the whole Buffonian, emphasising the direct action of the changeful milieu. "Species vary with their environment, and existing species have descended by modification from earlier and somewhat simpler species." He had a glimpse of the selection idea, and believed in mutations or sudden leaps--induced in the embryonic condition by external influences. The complete history of evolution-theories will include many instances of guesses at truth which were afterwards substantiated, thus the geographer von Buch (1773-1853) detected the importance of the Isolation factor on which Wagner, Romanes, Gulick and others have laid great stress, but we must content ourselves with recalling one other pioneer, the author of the Vestiges of Creation (1844), a work which passed through ten editions in nine years and certainly helped to harrow the soil for Darwin's sowing. As Darwin said, "it did excellent service in this country in calling attention to the subject, in removing prejudice, and in thus preparing the ground for the reception of analogous views."[24] Its author, Robert Chambers (1802-1871) was in part a Buffonian--maintaining that environment moulded organisms adaptively, and in part a Goethian--believing in an inherent progressive impulse which lifted organisms from one grade of organisation to another.

As Regards Natural Selection

The only thinker to whom Darwin was directly indebted, so far as the theory of Natural Selection is concerned, was Malthus, and we may once more quote the well-known passage in the Autobiography: "In October, 1838, that is, fifteen months after I had begun my systematic enquiry, I happened to read for amusement 'Malthus on Population,' and being well prepared to appreciate the struggle for existence which everywhere goes on from long-continued observation of the habits of animals and plants, it at once struck me that under these circumstances favourable variations would tend to be preserved, and unfavourable ones to be destroyed. The result of this would be the formation of new species."[25]

Although Malthus gives no adumbration of the idea of Natural Selection in his exposition of the eliminative processes which go on in mankind, the suggestive value of his essay is undeniable, as is strikingly borne out by the fact that it gave to Alfred Russel Wallace also "the long-sought clue to the effective agent in the evolution of organic species."[26] One day in Ternate when he was resting between fits of fever, something brought to his recollection the work of Malthus which he had read twelve years before. "I thought of his clear exposition of 'the positive checks to increase'--disease, accidents, war, and famine--which keep down the population of savage races to so much lower an average than that of more civilized peoples. It then occurred to me that these causes or their equivalents are continually acting in the case of animals also; and as animals usually breed much more rapidly than does mankind, the destruction every year from these causes must be enormous in order to keep down the numbers of each species, since they evidently do not increase regularly from year to year, as otherwise the world would long ago have been densely crowded with those that breed most quickly. Vaguely thinking over the enormous and constant destruction which this implied, it occurred to me to ask the question, Why do some die and some live? And the answer was clearly, that on the whole the best fitted live. From the effects of disease the most healthy escaped; from enemies the strongest, the swiftest, or the most cunning; from famine the best hunters or those with the best digestion; and so on. Then it suddenly flashed upon me that this self-acting process would necessarily improve the race, because in every generation the inferior would inevitably be killed off and the superior would remain--that is, the fittest would survive."[27] We need not apologise for this long quotation, it is a tribute to Darwin's magnanimous colleague, the Nestor of the evolutionist camp,--and it probably indicates the line of thought which Darwin himself followed. It is interesting also to recall the fact that in 1852, when Herbert Spencer wrote his famous Leader article on "The Development Hypothesis" in which he argued powerfully for the thesis that the whole animate world is the result of an age-long process of natural transformation, he wrote for The Westminster Review another important essay, "A Theory of Population deduced from the General Law of Animal Fertility," towards the close of which he came within an ace of recognising that the struggle for existence was a factor in organic evolution. At a time when pressure of population was practically interesting men's minds, Darwin, Wallace, and Spencer were being independently led from a social problem to a biological theory. There could be no better illustration, as Prof. Patrick Geddes has pointed out, of the Comtian thesis that science is a "social phenomenon."

Therefore, as far more important than any further ferreting out of vague hints of Natural Selection in books which Darwin never read, we would indicate by a quotation the view that the central idea in Darwinism is correlated with contemporary social evolution. "The substitution of Darwin for Paley as the chief interpreter of the order of nature is currently regarded as the displacement of an anthropomorphic view by a purely scientific one: a little reflection, however, will show that what has actually happened has been merely the replacement of the anthropomorphism of the eighteenth century by that of the nineteenth. For the place vacated by Paley's theological and metaphysical explanation has simply been occupied by that suggested to Darwin and Wallace by Malthus in terms of the prevalent severity of industrial competition, and those phenomena of the struggle for existence which the light of contemporary economic theory has enabled us to discern, have thus come to be temporarily exalted into a complete explanation of organic progress."[28] It goes without saying that the idea suggested by Malthus was developed by Darwin into a biological theory which was then painstakingly verified by being used as an interpretative formula, and that the validity of a theory so established is not affected by what suggested it, but the practical question which this line of thought raises in the mind is this: if Biology did thus borrow with such splendid results from social theory, why should we not more deliberately repeat the experiment?

Darwin was characteristically frank and generous in admitting that the principle of Natural Selection had been independently recognised by Dr. W. C. Wells in 1813 and by Mr. Patrick Matthew in 1831, but he had no knowledge of these anticipations when he published the first edition of The Origin of Species. Wells, whose "Essay on Dew" is still remembered, read in 1813 before the Royal Society a short paper entitled "An Account of a White Female, part of whose skin resembles that of a Negro" (published in 1818). In this communication, as Darwin said, "he observes, firstly, that all animals tend to vary in some degree, and, secondly, that agriculturists improve their domesticated animals by selection; and then, he adds, but what is done in this latter case 'by art, seems to be done with equal efficacy, though more slowly, by nature, in the formation of varieties of mankind, fitted for the country which they inhabit.'"[29] Thus Wells had the clear idea of survival dependent upon a favourable variation, but he makes no more use of the idea and applies it only to man. There is not in the paper the least hint that the author ever thought of generalising the remarkable sentence quoted above.

Of Mr. Patrick Matthew, who buried his treasure in an appendix to a work on Naval Timber and Arboriculture, Darwin said that "he clearly saw the full force of the principle of natural selection." In 1860 Darwin wrote--very characteristically--about this to Lyell: "Mr. Patrick Matthew publishes a long extract from his work on Naval Timber and Arboriculture, published in 1831, in which he briefly but completely anticipates the theory of Natural Selection.

I have ordered the book, as some passages are rather obscure, but it is certainly, I think, a complete but not developed anticipation. Erasmus always said that surely this would be shown to be the case some day. Anyhow, one may be excused in not having discovered the fact in a work on Naval Timber."[30]

De Quatrefages and De Varigny have maintained that the botanist Naudin stated the theory of evolution by natural selection in 1852. He explains very clearly the process of artificial selection, and says that in the garden we are following Nature's method. "We do not think that Nature has made her species in a different fashion from that in which we proceed ourselves in order to make our variations." But, as Darwin said, "he does not show how selection acts under nature." Similarly it must be noted in regard to several pre-Darwinian pictures of the struggle for existence (such as Herder's, who wrote in 1790 "All is in struggle ... each one for himself" and so on), that a recognition of this is only the first step in Darwinism.

Profs. E. Perrier and H. F. Osborn have called attention to a remarkable anticipation of the selection-idea which is to be found in the speculations of Geoffroy Saint-Hilaire (1825-1828) on the evolution of modern Crocodilians from the ancient Teleosaurs. Changing environment induced changes in the respiratory system and far-reaching consequences followed. The atmosphere, acting upon the pulmonary cells, brings about "modifications which are favourable or destructive ('funestes'); these are inherited, and they influence all the rest of the organisation of the animal because if these modifications lead to injurious effects the animals which exhibit them perish and are replaced by others of a somewhat different form, a form changed so as to be adapted to (?la convenance) the new environment."

Prof. E. B. Poulton[31] has shown that the anthropologist James Cowles Prichard (1786-1848) must be included even in spite of himself among the precursors of Darwin. In some passages of the second edition of his Researches into the Physical History of Mankind (1826), he certainly talks evolution and anticipates Prof. Weismann in denying the transmission of acquired characters. He is, however, sadly self-contradictory and his evolutionism weakens in subsequent editions--the only ones that Darwin saw. Prof. Poulton finds in Prichard's work a recognition of the operation of Natural Selection. "After inquiring how it is that 'these varieties are developed and preserved in connexion with particular climates and differences of local situation,' he gives the following very significant answer: 'One cause which tends to maintain this relation is obvious. Individuals and families, and even whole colonies perish and disappear in climates for which they are, by peculiarity of constitution, not adapted. Of this fact proofs have been already mentioned.'" Mr. Francis Darwin and Prof. A. C. Seward discuss Prichard's "anticipations" in More Letters of Charles Darwin, Vol. I. p. 43, and come to the conclusion that the evolutionary passages are entirely neutralised by others of an opposite trend. There is the same difficulty with Buffon.

Hints of the idea of Natural Selection have been detected elsewhere. James Watt,[32] for instance, has been reported as one of the anticipators (1851). But we need not prolong the inquiry further, since Darwin did not know of any anticipations until after he had published the immortal work of 1859, and since none of those who got hold of the idea made any use of it. What Darwin did was to follow the clue which Malthus gave him, to realise, first by genius and afterwards by patience, how the complex and subtle struggle for existence works out a natural selection of those organisms which vary in the direction of fitter adaptation to the conditions of their life. So much success attended his application of the Selection-formula that for a time he regarded Natural Selection as almost the sole factor in evolution, variations being pre-supposed; gradually, however, he came to recognise that there was some validity in the factors which had been emphasised by Lamarck and by Buffon, and in his well known summing up in the sixth edition of the Origin he says of the transformation of species: "This has been effected chiefly through the natural selection of numerous successive, slight, favourable variations; aided in an important manner by the inherited effects of the use and disuse of parts; and in an unimportant manner, that is, in relation to adaptive structures, whether past or present, by the direct action of external conditions, and by variations which seem to us in our ignorance to arise spontaneously."

To sum up: the idea of organic evolution, older than Aristotle, slowly developed from the stage of suggestion to the stage of verification, and the first convincing verification was Darwin's; from being an a priori anticipation it has become an interpretation of nature, and Darwin is still the chief interpreter; from being a modal interpretation it has advanced to the rank of a causal theory, the most convincing part of which men will never cease to call Darwinism.

FOOTNOTES:

[Footnote 1: Columbia University Biological Series, Vol. I. New York and London, 1894. We must acknowledge our great indebtedness to this fine piece of work.]

[Footnote 2: op. cit. p. 41.]

[Footnote 3: See G. J. Romanes, "Aristotle as a Naturalist," Contemporary Review, Vol. lix. p. 275, 1891; G. Pouchet, La Biologie Aristotique, Paris, 1885; E. Zeller, A History of Greek Philosophy, London, 1881, and "Ueber die griechischen Vorgger Darwin's," Abhandl. Berlin Akad. 1878, pp. 111-124.]

[Footnote 4: op. cit. p. 81.]

[Footnote 5: op. cit. p. 87.]

[Footnote 6: See Brock, "Die Stellung Kant's zur Deszendenztheorie," Biol. Centralbl. viii. 1889, pp. 641-648. Fritz Schultze, Kant und Darwin, Jena, 1875.]

[Footnote 7: Mr. Alfred Russel Wallace writes: "We claim for Darwin that he is the Newton of natural history, and that, just so surely as that the discovery and demonstration by Newton of the law of gravitation established order in place of chaos and laid a sure foundation for all future study of the starry heavens, so surely has Darwin, by his discovery of the law of natural selection and his demonstration of the great principle of the preservation of useful variations in the struggle for life, not only thrown a flood of light on the process of development of the whole organic world, but also established a firm foundation for all future study of nature" (Darwinism, London, 1889, p. 9). See also Prof. Karl Pearson's Grammar of Science (2nd edit.), London, 1900, p. 32. See Osborn, op. cit. p. 100.]

[Footnote 8: Experimental Evolution. London, 1892. Chap. I. p. 14.]

[Footnote 9: See J. Arthur Thomson, The Science of Life. London, 1899, Chap. XVI. "Evolution of Evolution Theory."]

[Footnote 10: See Carus Sterne (Ernst Krause), Die allgemeine Weltanschauung in ihrer historischen Entwickelung. Stuttgart, 1889.

Chapter entitled

"Best digkeit oder Verderlichkeit der Naturwesen."]

[Footnote 11: Zoonomia, or the Laws of Organic Life, 2 vols. London, 1794; Osborn, op. cit. p. 145.]

[Footnote 12: See Alpheus S. Packard, Lamarck, the Founder of Evolution, His Life and Work, with Translations of his writings on Organic Evolution. London, 1901.]

[Footnote 13: See Edward Clodd, Pioneers of Evolution, London, p. 161, 1897.]

[Footnote 14: See Chapter ix. "The Genetic View of Nature" in J. T. Merz's History of European Thought in the Nineteenth Century, Vol. 2, Edinburgh and London, 1903.]

[Footnote 15: See Prof. W. A. Locy's Biology and its Makers. New York, 1908. Part II. "The Doctrine of Organic Evolution."]

[Footnote 16: Presidential Address to the British Association meeting at Dublin in 1908.]

[Footnote 17: See in particular Samuel Butler, Evolution Old and New, London, 1879; J. L. de Lanessan, "Buffon et Darwin," Revue Scientifique, XLIII. pp. 385-391, 425-432, 1889.]

[Footnote 18: op. cit. p. 136.]

[Footnote 19: See Ernest Krause and Charles Darwin, Erasmus Darwin, London, 1879.]

[Footnote 20: Osborn, op. cit. p. 142.]

[Footnote 21: See E. Perrier, La Philosophie Zoologique avant Darwin, Paris, 1884; A. de Quatrefages, Darwin et ses Pr 閏 urseurs Fran 鏑 is, Paris, 1870; Packard, op. cit.; also Claus, Lamarck als Begrder der Descendenzlehre, Wien, 1888; Haeckel, Natural History of Creation, Eng. transl. London, 1879; Lang, Zur Charakteristik der Forschungswege von Lamarck und Darwin, Jena, 1889.]

[Footnote 22: See Huxley's article "Evolution in Biology," Encyclopaedia Britannica (9th edit.), 1879, pp. 744-751, and Sully's article, "Evolution in Philosophy," ibid. pp. 751-772.]

[Footnote 23: See Haeckel, Die Naturanschauung von Darwin, Goethe und Lamarck, Jena, 1882.]

[Footnote 24: Origin of Species (6th edit.), p. xvii.]

[Footnote 25: The Life and Letters of Charles Darwin, Vol. 1. p. 83. London, 1887.]

[Footnote 26: A. R. Wallace, My Life, a Record of Events and Opinions, London, 1905, Vol. 1, p. 232.]

[Footnote 27: My Life, Vol. 1. p. 361.]

[Footnote 28: P. Geddes. article "Biology." Chambers's Encyclopaedia.]

[Footnote 29: Origin of Species (6th edit.), p. xv.]

[Footnote 30: Life and Letters, II, p. 301.]

[Footnote 31: Science Progress, New Series, Vol. 1. 1897. "A Remarkable Anticipation of Modern Views on Evolution." See also Chap. VI. in Essays on Evolution, Oxford, 1908.]

[Footnote 32: See Prof. Patrick Geddes's article "Variation and Selection," Encyclopaedia Britannica (9th edit.) 1888.]

THE SELECTION THEORY

BY AUGUST WEISMANN

Professor of Zoology in the University of Freiburg (Baden)

I. THE IDEA OF SELECTION

Many and diverse were the discoveries made by Charles Darwin in the course of a long and strenuous life, but none of them has had so far-reaching an influence on the science and thought of his time as the theory of selection. I do not believe that the theory of evolution would have made its way so easily and so quickly after Darwin took up the cudgels in favour of it if he had not been able to support it by a principle which was capable of solving, in a simple manner, the greatest riddle that living nature presents to us,--I mean the purposiveness of every living form relative to the conditions of its life and its marvellously exact adaptation to these.

Everyone knows that Darwin was not alone in discovering the principle of selection, and that the same idea occurred simultaneously and independently to Alfred Russel Wallace. At the memorable meeting of the Linnean Society on 1st July, 1858, two papers were read (communicated by Lyell and Hooker) both setting forth the same idea of selection. One was written by Charles Darwin in Kent, the other by Alfred Wallace in Ternate, in the Malay Archipelago. It was a splendid proof of the magnanimity of these two investigators, that they thus in all friendliness and without envy, united in laying their ideas before a scientific tribunal: their names will always shine side by side as two of the brightest stars in the scientific sky.

The idea of selection set forth by the two naturalists was at the time absolutely new, but it was also so simple that Huxley could say of it later, "How extremely stupid not to have thought of that." As Darwin was led to the general doctrine of descent, not through the labours of his predecessors in the early years of the century, but by his own observations, so it was in regard to the principle of selection. He was struck by the innumerable cases of adaptation, as, for instance, that of the woodpeckers and tree-frogs to climbing, or the hooks and feather-like appendages of seeds, which aid in the distribution of plants, and he said to himself that an explanation of adaptations was the first thing to be sought for in attempting to formulate a theory of evolution.

But since adaptations point to changes which have been undergone by the ancestral forms of existing species, it is necessary, first of all, to inquire how far species in general are variable. Thus Darwin's attention was directed in the first place to the phenomenon of variability, and the use man has made of this, from very early times, in the breeding of his domesticated animals and cultivated plants. He inquired carefully how breeders set to work, when they wished to modify the structure and appearance of a species to their own ends, and it was soon clear to him that selection for breeding purposes played the chief part.

But how was it possible that such processes should occur in free nature? Who is here the breeder, making the selection, choosing out one individual to bring forth offspring and rejecting others? That was the problem that for a long time remained a riddle to him.

Darwin himself relates how illumination suddenly came to him. He had been reading, for his own pleasure, Malthus' book on Population, and, as he had long known from numerous observations, that every species gives rise to many more descendants than ever attain to maturity, and that, therefore, the greater number of the descendants of a species perish without reproducing, the idea came to him that the decision as to which member of a species was to perish and which was to attain to maturity and reproduction might not be a matter of chance, but might be determined by the constitution of the individuals themselves, according as they were more or less fitted for survival. With this idea the foundation of the theory of selection was laid.

In artificial selection the breeder chooses out for pairing only such individuals as possess the character desired by him in a somewhat higher degree than the rest of the race. Some of the descendants inherit this character, often in a still higher degree, and if this method be pursued throughout several generations, the race is transformed in respect of that particular character.

Natural selection depends on the same three factors as artificial selection:

on variability, inheritance, and selection for breeding, but this last is here carried out not by a breeder but by what Darwin called the "struggle for existence." This last factor is one of the special features of the Darwinian conception of nature. That there are carnivorous animals which take heavy toll in every generation of the progeny of the animals on which they prey, and that there are herbivores which decimate the plants in every generation had long been known, but it is only since Darwin's time that sufficient attention has been paid to the facts that, in addition to this regular destruction, there exists between the members of a species a keen competition for space and food, which limits multiplication, and that numerous individuals of each species perish because of unfavourable climatic conditions. The "struggle for existence," which Darwin regarded as taking the place of the human breeder in free nature, is not a direct struggle between carnivores and their prey, but is the assumed competition for survival between individuals of the same species, of which, on an average, only those survive to reproduce which have the greatest power of resistance, while the others, less favourably constituted, perish early. This struggle is so keen, that, within a limited area, where the conditions of life have long remained unchanged, of every species, whatever be the degree of fertility, only two, on an average, of the descendants of each pair survive; the others succumb either to enemies, or to disadvantages of climate, or to accident. A high degree of fertility is thus not an indication of the special success of a species, but of the numerous dangers that have attended its evolution. Of the six young brought forth by a pair of elephants in the course of their lives only two survive in a given area; similarly, of the millions of eggs which two thread-worms leave behind them only two survive. It is thus possible to estimate the dangers which threaten a species by its ratio of elimination, or, since this cannot be done directly, by its fertility.

Although a great number of the descendants of each generation fall victims to accident, among those that remain it is still the greater or less fitness of the organism that determines the "selection for breeding purposes," and it would be incomprehensible if, in this competition, it were not ultimately, that is, on an average, the best equipped which survive, in the sense of living long enough to reproduce.

Thus the principle of natural selection is the selection of the best for reproduction, whether the "best" refers to the whole constitution, to one or more parts of the organism, or to one or more stages of development. Every organ, every part, every character of an animal, fertility and intelligence included, must be improved in this manner, and be gradually brought up in the course of generations to its highest attainable state of perfection. And not only may improvement of parts be brought about in this way, but new parts and organs may arise, since, through the slow and minute steps of individual or "fluctuating" variations, a part may be added here or dropped out there, and thus something new is produced.

The principle of selection solved the riddle as to how what was purposive could conceivably be brought about without the intervention of a directing power, the riddle which animate nature presents to our intelligence at every turn, and in face of which the mind of a Kant could find no way out, for he regarded a solution of it as not to be hoped for. For, even if we were to assume an evolutionary force that is continually transforming the most primitive and the simplest forms of life into ever higher forms, and the homogeneity of primitive times into the infinite variety of the present, we should still be unable to infer from this alone how each of the numberless forms adapted to particular conditions of life should have appeared precisely at the right moment in the history of the earth to which their adaptations were appropriate, and precisely at the proper place in which all the conditions of life to which they were adapted occurred: the humming-birds at the same time as the flowers; the trichina at the same time as the pig; the bark-coloured moth at the same time as the oak, and the wasp-like moth at the same time as the wasp which protects it. Without processes of selection we should be obliged to assume a "pre-established harmony" after the famous Leibnitzian model, by means of which the clock of the evolution of organisms is so regulated as to strike in exact synchronism with that of the history of the earth! All forms of life are strictly adapted to the conditions of their life, and can persist under these conditions alone.

There must therefore be an intrinsic connection between the conditions and the structural adaptations of the organism, and, since the conditions of life cannot be determined by the animal itself, the adaptations must be called forth by the conditions.

The selection theory teaches us how this is conceivable, since it enables us to understand that there is a continual production of what is non-purposive as well as of what is purposive, but the purposive alone survives, while the non-purposive perishes in the very act of arising. This is the old wisdom taught long ago by Empedocles.

II. THE LAMARCKIAN PRINCIPLE

Lamarck, as is well known, formulated a definite theory of evolution at the beginning of the nineteenth century, exactly fifty years before the Darwin-Wallace principle of selection was given to the world. This brilliant investigator also endeavoured to support his theory by demonstrating forces which might have brought about the transformations of the organic world in the course of the ages. In addition to other factors, he laid special emphasis on the increased or diminished use of the parts of the body, assuming that the strengthening or weakening which takes place from this cause during the individual life, could be handed on to the offspring, and thus intensified and raised to the rank of a specific character. Darwin also regarded this Lamarckian principle, as it is now generally called, as a factor in evolution, but he was not fully convinced of the transmissibility of acquired characters.

As I have here to deal only with the theory of selection, I need not discuss the Lamarckian hypothesis, but I must express my opinion that there is room for much doubt as to the cooperation of this principle in evolution. Not only is it difficult to imagine how the transmission of functional modifications could take place, but, up to the present time, notwithstanding the endeavours of many excellent investigators, not a single actual proof of such inheritance has been brought forward. Semon's experiments on plants are, according to the botanist Pfeffer, not to be relied on, and even the recent, beautiful experiments made by Dr. Kammerer on salamanders, cannot, as I hope to show elsewhere, be regarded as proof, if only because they do not deal at all with functional modifications, that is, with modifications brought about by use, and it is to these alone that the Lamarckian principle refers.

III. OBJECTIONS TO THE THEORY OF SELECTION

(a) Saltatory evolution

The Darwinian doctrine of evolution depends essentially on the cumulative augmentation of minute variations in the direction of utility. But can such minute variations, which are undoubtedly continually appearing among the individuals of the same species, possess any selection-value; can they determine which individuals are to survive, and which are to succumb; can they be increased by natural selection till they attain to the highest development of a purposive variation?

To many this seems so improbable that they have urged a theory of evolution by leaps from species to species. Kliker, in 1872, compared the evolution of species with the processes which we can observe in the individual life in cases of alternation of generations. But a polyp only gives rise to a medusa because it has itself arisen from one, and there can be no question of a medusa ever having arisen suddenly and de novo from a polyp-bud, if only because both forms are adapted in their structure as a whole, and in every detail to the conditions of their life. A sudden origin, in a natural way, of numerous adaptations is inconceivable. Even the degeneration of a medusoid from a free-swimming animal to a mere brood-sac (gonophore) is not sudden and saltatory, but occurs by imperceptible modifications throughout hundreds of years, as we can learn from the numerous stages of the process of degeneration persisting at the same time in different species.

If, then, the degeneration to a simple brood-sac takes place only by very slow transitions, each stage of which may last for centuries, how could the much more complex ascending evolution possibly have taken place by sudden leaps? I regard this argument as capable of further extension, for wherever in nature we come upon degeneration, it is taking place by minute steps and with a slowness that makes it not directly perceptible, and I believe that this in itself justifies us in concluding that the same must be true of ascending evolution. But in the latter case the goal can seldom be distinctly recognised while in cases of degeneration the starting-point of the process can often be inferred, because several nearly related species may represent different stages.

In recent years Bateson in particular has championed the idea of saltatory, or so-called discontinuous evolution, and has collected a number of cases in which more or less marked variations have suddenly appeared. These are taken for the most part from among domesticated animals which have been bred and crossed for a long time, and it is hardly to be wondered at that their much mixed and much influenced germ-plasm should, under certain conditions, give rise to remarkable phenomena, often indeed producing forms which are strongly suggestive of monstrosities, and which would undoubtedly not survive in free nature, unprotected by man. I should regard such cases as due to an intensified germinal selection--though this is to anticipate a little--and from this point of view it cannot be denied that they have a special interest. But they seem to me to have no significance as far as the transformation of species is concerned, if only because of the extreme rarity of their occurrence.

There are, however, many variations which have appeared in a sudden and saltatory manner, and some of these Darwin pointed out and discussed in detail: the copper beech, the weeping trees, the oak with "fern-like leaves," certain garden-flowers, etc. But none of them have persisted in free nature, or evolved into permanent types.

On the other hand, wherever enduring types have arisen, we find traces of a gradual origin by successive stages, even if, at first sight, their origin may appear to have been sudden. This is the case with seasonal Dimorphism, the first known cases of which exhibited marked differences between the two generations, the winter and the summer brood. Take for instance the much discussed and studied form Vanessa (Araschnia) levana-prorsa. Here the differences between the two forms are so great and so apparently disconnected, that one might almost believe it to be a sudden mutation, were it not that old transition-stages can be called forth by particular temperatures, and we know other butterflies, as for instance our Garden Whites, in which the differences between the two generations are not nearly so marked; indeed, they are so little apparent that they are scarcely likely to be noticed except by experts. Thus here again there are small initial steps, some of which, indeed, must be regarded as adaptations, such as the green-sprinkled or lightly tinted under-surface which gives them a deceptive resemblance to parsley or to Cardamine leaves.

Even if saltatory variations do occur, we cannot assume that these have ever led to forms which are capable of survival under the conditions of wild life. Experience has shown that in plants which have suddenly varied the power of persistence is diminished. Korschinsky attributes to them weaknesses of organisation in general; "they bloom late, ripen few of their seeds, and show great sensitiveness to cold." These are not the characters which make for success in the struggle for existence.

We must briefly refer here to the views--much discussed in the last decade-- of H. de Vries, who believes that the roots of transformation must be sought for in saltatory variations arising from internal causes, and distinguishes such mutations, as he has called them, from ordinary individual variations, in that they breed true, that is, with strict in-breeding they are handed on pure to the next generation. I have elsewhere endeavoured to point out the weaknesses of this theory,[33] and I am the less inclined to return to it here that it now appears[34] that the far-reaching conclusions drawn by de Vries from his observations on the Evening Primrose, Oenothera lamarckiana, rest upon a very insecure foundation. The plant from which de Vries saw numerous "species"--his "mutations"--arise was not, as he assumed, a wild species that had been introduced to Europe from America, but was probably a hybrid form which was first discovered in the Jardin des Plantes in Paris, and which does not appear to exist anywhere in America as a wild species.

This gives a severe shock to the "Mutation theory," for the other actually wild species with which de Vries experimented showed no "mutations" but yielded only negative results.

Thus we come to the conclusion that Darwin[35] was right in regarding transformations as taking place by minute steps, which, if useful, are augmented in the course of innumerable generations, because their possessors more frequently survive in the struggle for existence.

(b) Selection-value of the initial steps

Is it possible that the insignificant deviations which we know as "individual variations" can form the beginning of a process of selection? Can they decide which is to perish and which to survive? To use a phrase of Romanes, can they have selection-value?

Darwin himself answered this question, and brought together many excellent examples to show that differences, apparently insignificant because very small, might be of decisive importance for the life of the possessor. But it is by no means enough to bring forward cases of this kind, for the question is not merely whether finished adaptations have selection-value, but whether the first beginnings of these, and whether the small, I might almost say minimal increments, which have led up from these beginnings to the perfect adaptation, have also had selection-value. To this question even one who, like myself, has been for many years a convinced adherent of the theory of selection, can only reply: We must assume so, but we cannot prove it in any case. It is not upon demonstrative evidence that we rely when we champion the doctrine of selection as a scientific truth; we base our argument on quite other grounds. Undoubtedly there are many apparently insignificant features, which can nevertheless be shown to be adaptations--for instance, the thickness of the basin-shaped shell of the limpets that live among the breakers on the shore. There can be no doubt that the thickness of these shells, combined with their flat forms, protects the animals from the force of the waves breaking upon them,--but how have they become so thick? What proportion of thickness was sufficient to decide that of two variants of a limpet one should survive, the other be eliminated? We can say nothing more than that we infer from the present state of the shell, that it must have varied in regard to differences in shell-thickness, and that these differences must have had selection-value,--no proof therefore, but an assumption which we must show to be convincing.

For a long time the marvellously complex radiate and lattice-work skeletons of Radiolarians were regarded as a mere outflow of "Nature's infinite wealth of form," as an instance of a purely morphological character with no biological significance. But recent investigations have shown that these, too, have an adaptive significance (Haker). The same thing has been shown by Scht in regard to the lowly unicellular plants, the Peridineae, which abound alike on the surface of the ocean and in its depths. It has been shown that the long skeletal processes which grow out from these organisms have significance not merely as a supporting skeleton, but also as an extension of the superficial area, which increases the contact with the water-particles, and prevents the floating organisms from sinking. It has been established that the processes are considerably shorter in the colder layers of the ocean, and that they may be twelve times as long[36] in the warmer layers, thus corresponding to the greater or smaller amount of friction which takes place in the denser and less dense layers of the water.

The Peridineae of the warmer ocean layers have thus become long-rayed, those of the colder layers short-rayed, not through the direct effect of friction on the protoplasm, but through processes of selection, which favoured the longer rays in warm water, since they kept the organism afloat, while those with short rays sank and were eliminated. If we put the question as to selection-value in this case, and ask how great the variations in the length of processes must be in order to possess selection-value; what can we answer except that these variations must have been minimal, and yet sufficient to prevent too rapid sinking and consequent elimination? Yet this very case would give the ideal opportunity for a mathematical calculation of the minimal selection-value, although of course it is not feasible from lack of data to carry out the actual calculation.

But even in organisms of more than microscopic size there must frequently be minute, even microscopic differences which set going the process of selection, and regulate its progress to the highest possible perfection.

Many tropical trees possess thick, leathery leaves, as a protection against the force of the tropical raindrops. The direct influence of the rain cannot be the cause of this power of resistance, for the leaves, while they were still thin, would simply have been torn to pieces. Their toughness must therefore be referred to selection, which would favour the trees with slightly thicker leaves, though we cannot calculate with any exactness how great the first stages of increase in thickness must have been. Our hypothesis receives further support from the fact that, in many such trees, the leaves are drawn out into a beak-like prolongation (Stahl and Haberlandt) which facilitates the rapid falling off of the rain water, and also from the fact that the leaves, while they are still young, hang limply down in bunches which offer the least possible resistance to the rain. Thus there are here three adaptations which can only be interpreted as due to selection. The initial stages of these adaptations must undoubtedly have had selection-value.

But even in regard to this case we are reasoning in a circle, not giving "proofs," and no one who does not wish to believe in the selection-value of the initial stages can be forced to do so. Among the many pieces of presumptive evidence a particularly weighty one seems to me to be the smallness of the steps of progress which we can observe in certain cases, as for instance in leaf-imitation among butterflies, and in mimicry generally. The resemblance to a leaf, for instance of a particular Kallima, seems to us so close as to be deceptive, and yet we find in another individual, or it may be in many others, a spot added which increases the resemblance, and which could not have become fixed unless the increased deceptiveness so produced had frequently led to the overlooking of its much persecuted possessor. But if we take the selection-value of the initial stages for granted, we are confronted with the further question which I myself formulated many years ago: How does it happen that the necessary beginnings of a useful variation are always present? How could insects which live upon or among green leaves become all green, while those that live on bark become brown? How have the desert animals become yellow and the Arctic animals white? Why were the necessary variations always present? How could the green locust lay brown eggs, or the privet caterpillar develop white and lilac-coloured lines on its green skin?

It is of no use answering to this that the question is wrongly formulated[37] and that it is the converse that is true; that the process of selection takes place in accordance with the variations that present themselves. This proposition is undeniably true, but so also is another, which apparently negatives it: the variation required has in the majority of cases actually presented itself. Selection cannot solve this contradiction; it does not call forth the useful variation, but simply works upon it. The ultimate reason why one and the same insect should occur in green and in brown, as often happens in caterpillars and locusts, lies in the fact that variations towards brown presented themselves, and so also did variations towards green: the kernel of the riddle lies in the varying, and for the present we can only say, that small variations in different directions present themselves in every species. Otherwise so many different kinds of variations could not have arisen. I have endeavoured to explain this remarkable fact by means of the intimate processes that must take place within the germ-plasm, and I shall return to the problem when dealing with "germinal selection."

We have, however, to make still greater demands on variation, for it is not enough that the necessary variation should occur in isolated individuals, because in that case there would be small prospect of its being preserved, notwithstanding its utility. Darwin at first believed, that even single variations might lead to transformation of the species, but later he became convinced that this was impossible, at least without the cooperation of other factors, such as isolation and sexual selection.

In the case of the green caterpillars with bright longitudinal stripes, numerous individuals exhibiting this useful variation must have been produced to start with. In all higher, that is, multicellular organisms, the germ-substance is the source of all transmissible variations, and this germ-plasm is not a simple substance but is made up of many primary constituents. The question can therefore be more precisely stated thus: How does it come about that in so many cases the useful variations present themselves in numbers just where they are required, the white oblique lines in the leaf-caterpillar on the under surface of the body, the accompanying coloured stripes just above them? And, further, how has it come about that in grass caterpillars, not oblique but longitudinal stripes, which are more effective for concealment among grass and plants, have been evolved? And finally, how is it that the same Hawk-moth caterpillars, which to-day show oblique stripes, possessed longitudinal stripes in Tertiary times? We can read this fact from the history of their development, and I have before attempted to show the biological significance of this change of colour.[38]

For the present I need only draw the conclusion that one and the same caterpillar may exhibit the initial stages of both, and that it depends on the manner in which these marking elements are intensified and combined by natural selection whether whitish longitudinal or oblique stripes should result. In this case then the "useful variations" were actually "always there," and we see that in the same group of Lepidoptera, e.g. species of Sphingidae, evolution has occurred in both directions according to whether the form lived among grass or on broad leaves with oblique lateral veins, and we can observe even now that the species with oblique stripes have longitudinal stripes when young, that is to say, while the stripes have no biological significance. The white places in the skin which gave rise, probably first as small spots, to this protective marking could be combined in one way or another according to the requirements of the species. They must therefore either have possessed selection-value from the first, or, if this was not the case at their earliest occurrence, there must have been some other factors which raised them to the point of selection-value. I shall return to this in discussing germinal selection. But the case may be followed still farther, and leads us to the same alternative on a still more secure basis.

Many years ago I observed in caterpillars of Smerinthus populi (the poplar hawk-moth), which also possess white oblique stripes, that certain individuals showed red spots above these stripes; these spots occurred only on certain segments, and never flowed together to form continuous stripes. In another species (Smerinthus tiliae) similar blood-red spots unite to form a line-like coloured seam in the last stage of larval life, while in S. ocellata rust-red spots appear in individual caterpillars, but more rarely than in S. populi, and they show no tendency to flow together.

Thus we have here the origin of a new character, arising from small beginnings, at least in S. tiliae, in which species the coloured stripes are a normal specific character. In the other species, S. populi and S. ocellata, we find the beginnings of the same variation, in one more rarely than in the other, and we can imagine that, in the course of time, in these two species, coloured lines over the oblique stripes will arise. In any case these spots are the elements of variation, out of which coloured lines may be evolved, if they are combined in this direction through the agency of natural selection. In S. populi the spots are often small, but sometimes it seems as though several had united to form large spots. Whether a process of selection in this direction will arise in S. populi and S. ocellata, or whether it is now going on cannot be determined, since we cannot tell in advance what biological value the marking might have for these two species. It is conceivable that the spots may have no selection-value as far as these species are concerned, and may therefore disappear again in the course of phylogeny, or, on the other hand, that they may be changed in another direction, for instance towards imitation of the rust-red fungoid patches on poplar and willow leaves. In any case we may regard the smallest spots as the initial stages of variation, the larger as a cumulative summation of these. Therefore either these initial stages must already possess selection-value, or, as I said before: There must be some other reason for their cumulative summation. I should like to give one more example, in which we can infer, though we cannot directly observe, the initial stages.

All the Holothurians or sea-cucumbers have in the skin calcereous bodies of different forms, usually thick and irregular, which make the skin tough and resistant. In a small group of them--the species of Synapta--the calcareous bodies occur in the form of delicate anchors of microscopic size. Up till 1897 these anchors, like many other delicate microscopic structures, were regarded as curiosities, as natural marvels. But a Swedish observer, Oestergren, has recently shown that they have a biological significance: they serve the footless Synapta as auxiliary organs of locomotion, since, when the body swells up in the act of creeping, they press firmly with their tips, which are embedded in the skin, against the substratum on which the animal creeps, and thus prevent slipping backwards. In other Holothurians this slipping is made impossible by the fixing of the tube-feet. The anchors act automatically, sinking their tips towards the ground when the corresponding part of the body thickens, and returning to the original position at an angle of 45 degrees to the upper surface when the part becomes thin again. The arms of the anchor do not lie in the same plane as the shaft, and thus the curve of the arms forms the outermost part of the anchor, and offers no further resistance to the gliding of the animal. Every detail of the anchor, the curved portion, the little teeth at the head, the arms, etc., can be interpreted in the most beautiful way, above all the form of the anchor itself, for the two arms prevent it from swaying round to the side. The position of the anchors, too, is definite and significant; they lie obliquely to the longitudinal axis of the animal, and therefore they act alike whether the animal is creeping backwards or forwards. Moreover, the tips would pierce through the skin if the anchors lay in the longitudinal direction. Synapta burrows in the sand; it first pushes in the thin anterior end, and thickens this again, thus enlarging the hole, then the anterior tentacles displace more sand, the body is worked in a little farther, and the process begins anew. In the first act the anchors are passive, but they begin to take an active share in the forward movement when the body is contracted again. Frequently the animal retains only the posterior end buried in the sand, and then the anchors keep it in position, and make rapid withdrawal possible.

Thus we have in these apparently random forms of the calcereous bodies, complex adaptations in which every little detail as to direction, curve, and pointing is exactly determined. That they have selection-value in their present perfected form is beyond all doubt, since the animals are enabled by means of them to bore rapidly into the ground and so to escape from enemies. We do not know what the initial stages were, but we cannot doubt that the little improvements, which occurred as variations of the originally simple slimy bodies of the Holothurians, were preserved because they already possessed selection-value for the Synaptidae. For such minute microscopic structures whose form is so delicately adapted to the role they have to play in the life of the animal, cannot have arisen suddenly and as a whole, and every new variation of the anchor, that is, in the direction of the development of the two arms, and every curving of the shaft which prevented the tips from projecting at the wrong time, in short, every little adaptation in the modelling of the anchor must have possessed selection-value. And that such minute changes of form fall within the sphere of fluctuating variations, that is to say, that they occur is beyond all doubt.

In many of the Synaptidae the anchors are replaced by calcareous rods bent in the form of an S, which are said to act in the same way. Others, such as those of the genus Ankyroderma, have anchors which project considerably beyond the skin, and, according to Oestergren, serve "to catch plant-particles and other substances" and so mask the animal. Thus we see that in the Synaptidae the thick and irregular calcareous bodies of the Holothurians have been modified and transformed in various ways in adaptation to the footlessness of these animals, and to the peculiar conditions of their life, and we must conclude that the earlier stages of these changes presented themselves to the processes of selection in the form of microscopic variations. For it is as impossible to think of any origin other than through selection in this case as in the case of the toughness, and the "drip-tips" of tropical leaves. And as these last could not have been produced directly by the beating of the heavy raindrops upon them, so the calcareous anchors of Synapta cannot have been produced directly by the friction of the sand and mud at the bottom of the sea, and, since they are parts whose function is passive the Lamarckian factor of use and disuse does not come into question. The conclusion is unavoidable, that the microscopically small variations of the calcareous bodies in the ancestral forms have been intensified and accumulated in a particular direction, till they have led to the formation of the anchor. Whether this has taken place by the action of natural selection alone, or whether the laws of variation and the intimate processes within the germ-plasm have cooperated will become clear in the discussion of germinal selection. This whole process of adaptation has obviously taken place within the time that has elapsed since this group of sea-cucumbers lost their tube-feet, those characteristic organs of locomotion which occur in no group except the Echinoderms, and yet have totally disappeared in the Synaptidae. And after all what would animals that live in sand and mud do with tube-feet?

(c) Coadaptation

Darwin pointed out that one of the essential differences between artificial and natural selection lies in the fact that the former can modify only a few characters, usually only one at a time, while Nature preserves in the struggle for existence all the variations of a species, at the same time and in a purely mechanical way, if they possess selection-value.

Herbert Spencer, though himself an adherent of the theory of selection, declared in the beginning of the nineties that in his opinion the range of this principle was greatly over-estimated, if the great changes which have taken place in so many organisms in the course of ages are to be interpreted as due to this process of selection alone, since no transformation of any importance can be evolved by itself; it is always accompanied by a host of secondary changes. He gives the familiar example of the Giant Stag of the Irish peat, the enormous antlers of which required not only a much stronger skull cap, but also greater strength of the sinews, muscles, nerves and bones of the whole anterior half of the animal, if their mass was not to weigh down the animal altogether. It is inconceivable, he says, that so many processes of selection should take place simultaneously, and we are therefore forced to fall back on the Lamarckian factor of the use and disuse of functional parts. And how, he asks, could natural selection follow two opposite directions of evolution in different parts of the body at the same time, as for instance in the case of the kangaroo, in which the forelegs must have become shorter, while the hind legs and the tail were becoming longer and stronger?

Spencer's main object was to substantiate the validity of the Lamarckian principle, the cooperation of which with selection had been doubted by many. And it does seem as though this principle, if it operates in nature at all, offers a ready and simple explanation of all such secondary variations. Not only muscles, but nerves, bones, sinews, in short all tissues which function actively, increase in strength in proportion as they are used, and conversely they decrease when the claims on them diminish. All the parts, therefore, which depend on the part that varied first, as for instance the enlarged antlers of the Irish Elk, must have been increased or decreased in strength, in exact proportion to the claims made upon them,--just as is actually the case.

But beautiful as this explanation would be, I regard it as untenable, because it assumes the transmissibility of functional modifications (so-called "acquired" characters), and this is not only undemonstrable, but is scarcely theoretically conceivable, for the secondary variations which accompany or follow the first as correlative variations, occur also in cases in which the animals concerned are sterile and therefore cannot transmit anything to their descendants. This is true of worker bees, and particularly of ants, and I shall here give a brief survey of the present state of the problem as it appears to me.

Much has been written on both sides of this question since the published controversy on the subject in the nineties between Herbert Spencer and myself. I should like to return to the matter in detail, if the space at my disposal permitted, because it seems to me that the arguments I advanced at that time are equally cogent to-day, notwithstanding all the objections that have since been urged against them. Moreover, the matter is by no means one of subordinate interest; it is the very kernel of the whole question of the reality and value of the principle of selection. For if selection alone does not suffice to explain "harmonious adaptation" as I have called Spencer's Coadaptation, and if we require to call in the aid of the Lamarckian factor it would be questionable whether selection would explain any adaptations whatever. In this particular case--of worker bees--the Lamarckian factor may be excluded altogether, for it can be demonstrated that here at any rate the effects of use and disuse cannot be transmitted.

But if it be asked why we are unwilling to admit the cooperation of the Darwinian factor of selection and the Lamarckian factor, since this would afford us an easy and satisfactory explanation of the phenomena, I answer: Because the Lamarckian principle is fallacious, and because by accepting it we close the way towards deeper insight. It is not a spirit of combativeness or a desire for self-vindication that induces me to take the field once more against the Lamarckian principle, it is the conviction that the progress of our knowledge is being obstructed by the acceptance of this fallacious principle, since the facile explanation it apparently affords prevents our seeking after a truer explanation and a deeper analysis.

The workers in the various species of ants are sterile, that is to say, they take no regular part in the reproduction of the species, although individuals among them may occasionally lay eggs. In addition to this they have lost the wings, and the receptaculum seminis, and their compound eyes have degenerated to a few facets. How could this last change have come about through disuse, since the eyes of workers are exposed to light in the same way as are those of the sexual insects and thus in this particular case are not liable to "disuse" at all? The same is true of the receptaculum seminis, which can only have been disused as far as its glandular portion and its stalk are concerned, and also of the wings, the nerves tracheae and epidermal cells of which could not cease to function until the whole wing had degenerated, for the chitinous skeleton of the wing does not function at all in the active sense.

But, on the other hand, the workers in all species have undergone modifications in a positive direction, as, for instance, the greater development of brain. In many species large workers have evolved,--the so-called soldiers, with enormous jaws and teeth, which defend the colony,--and in others there are small workers which have taken over other special functions, such as the rearing of the young Aphides. This kind of division of the workers into two castes occurs among several tropical species of ants, but it is also present in the Italian species, Colobopsis truncata. Beautifully as the size of the jaws could be explained as due to the increased use made of them by the "soldiers," or the enlarged brain as due to the mental activities of the workers, the fact of the infertility of these forms is an insurmountable obstacle to accepting such an explanation. Neither jaws nor brain can have been evolved on the Lamarckian principle.

The problem of coadaptation is no easier in the case of the ant than in the case of the Giant Stag. Darwin himself gave a pretty illustration to show how imposing the difference between the two kinds of workers in one species would seem if we translated it into human terms. In regard to the Driver ants (Anomma) we must picture to ourselves a piece of work, "for instance the building of a house, being carried on by two kinds of workers, of which one group was five feet four inches high, the other sixteen feet high."[39]

Although the ant is a small animal as compared with man or with the Irish Elk, the "soldier" with its relatively enormous jaws is hardly less heavily burdened than the Elk with its antlers, and in the ant's case, too, a strengthening of the skeleton, of the muscles, the nerves of the head, and of the legs must have taken place parallel with the enlargement of the jaws. Harmonious adaptation (coadaptation) has here been active in a high degree, and yet these "soldiers" are sterile! There thus remains nothing for it but to refer all their adaptations, positive and negative alike, to processes of selection which have taken place in the rudiments of the workers within the egg and sperm-cells of their parents. There is no way out of the difficulty except the one Darwin pointed out. He himself did not find the solution of the riddle at once. At first he believed that the case of the workers among social insects presented "the most serious special difficulty" in the way of his theory of natural selection; and it was only after it had become clear to him that it was not the sterile insects themselves but their parents that were selected, according as they produced more or less well adapted workers, that he was able to refer to this very case of the conditions among ants "in order to show the power of natural selection."[40] He explains his view by a simple but interesting illustration. Gardeners have produced, by means of long continued artificial selection, a variety of Stock, which bears entirely double, and therefore infertile flowers.[41] Nevertheless the variety continues to be reproduced from seed, because, in addition to the double and infertile flowers, the seeds always produce a certain number of single, fertile blossoms, and these are used to reproduce the double variety. These single and fertile plants correspond "to the males and females of an ant-colony, the infertile plants, which are regularly produced in large numbers, to the neuter workers of the colony."

This illustration is entirely apt, the only difference between the two cases consisting in the fact that the variation in the flower is not a useful, but a disadvantageous one, which can only be preserved by artificial selection on the part of the gardener, while the transformations that have taken place parallel with the sterility of the ants are useful, since they procure for the colony an advantage in the struggle for existence, and they are therefore preserved by natural selection. Even the sterility itself in this case is not disadvantageous, since the fertility of the true females has at the same time considerably increased. We may therefore regard the sterile forms of ants, which have gradually been adapted in several directions to varying functions, as a certain proof that selection really takes place in the germ-cells of the fathers and mothers of the workers, and that special complexes of primordia (ids) are present in the workers and in the males and females, and these complexes contain the primordia of the individual parts (determinants). But since all living entities vary, the determinants must also vary, now in a favourable, now in an unfavourable direction. If a female produces eggs, which contain favourably varying determinants in the worker-ids, then these eggs will give rise to workers modified in the favourable direction, and if this happens with many females, the colony concerned will contain a better kind of worker than other colonies.

I digress here in order to give an account of the intimate processes, which, according to my view, take place within the germ-plasm, and which I have called "germinal selection." These processes are of importance since they form the roots of variation, which in its turn is the root of natural selection. I cannot here do more than give a brief outline of the theory in order to show how the Darwin-Wallace theory of selection has gained support from it.

With others, I regard the minimal amount of substance which is contained within the nucleus of the germ-cells, in the form of rods, bands, or granules, as the germ-substance or germ-plasm, and I call the individual granules ids. There is always a multiplicity of such ids present in the nucleus, either occurring individually, or united in the form of rods or bands (chromosomes). Each id contains the primary constituents of a whole individual, so that several ids are concerned in the development of a new individual.

In every being of complex structure thousands of primary constituents must go to make up a single id; these I call determinants, and I mean by this name very small individual particles, far below the limits of microscopic visibility, vital units which feed, grow, and multiply by division. These determinants control the parts of the developing embryo,--in what manner need not here concern us. The determinants differ among themselves, those of a muscle are differently constituted from those of a nerve-cell or a glandular cell, etc., and every determinant is in its turn made up of minute vital units, which I call biophores, or the bearers of life. According to my view, these determinants not only assimilate, like every other living unit, but they vary in the course of their growth, as every living unit does; they may vary qualitatively if the elements of which they are composed vary, they may grow and divide more or less rapidly, and their variations give rise to corresponding variations of the organ, cell, or cell-group which they determine. That they are undergoing ceaseless fluctuations in regard to size and quality seems to me the inevitable consequence of their unequal nutrition; for although the germ-cell as a whole usually receives sufficient nutriment, minute fluctuations in the amount carried to different parts within the germ-plasm cannot fail to occur.

Now, if a determinant, for instance of a sensory cell, receives for a considerable time more abundant nutriment than before, it will grow more rapidly--become bigger, and divide more quickly, and, later, when the id concerned develops into an embryo, this sensory cell will become stronger than in the parents, possibly even twice as strong. This is an instance of a hereditary individual variation, arising from the germ.

The nutritive stream which, according to our hypothesis, favours the determinant N by chance, that is, for reasons unknown to us, may remain strong for a considerable time, or may decrease again; but even in the latter case it is conceivable that the ascending movement of the determinant may continue, because the strengthened determinant now actively nourishes itself more abundantly,--that is to say, it attracts the nutriment to itself, and to a certain extent withdraws it from its fellow-determinants. In this way, it may--as it seems to me--get into permanent upward movement, and attain a degree of strength from which there is no falling back. Then positive or negative selection sets in, favouring the variations which are advantageous, setting aside those which are disadvantageous.

In a similar manner a downward variation of the determinants may take place, if its progress be started by a diminished flow of nutriment. The determinants which are weakened by this diminished flow will have less affinity for attracting nutriment because of their diminished strength, and they will assimilate more feebly and grow more slowly, unless chance streams of nutriment help them to recover themselves. But, as will presently be shown, a change of direction cannot take place at every stage of the degenerative process. If a certain critical stage of downward progress be passed, even favourable conditions of food-supply will no longer suffice permanently to change the direction of the variation. Only two cases are conceivable; if the determinant corresponds to a useful organ, only its removal can bring back the germ-plasm to its former level; therefore personal selection removes the id in question, with its determinants, from the germ-plasm, by causing the elimination of the individual in the struggle for existence. But there is another conceivable case; the determinants concerned may be those of an organ which has become useless, and they will then continue unobstructed, but with exceeding slowness, along the downward path, until the organ becomes vestigial, and finally disappears altogether.

The fluctuations of the determinants hither and thither may thus be transformed into a lasting ascending or descending movement; and this is the crucial point of these germinal processes.

This is not a fantastic assumption; we can read it in the fact of the degeneration of disused parts. Useless organs are the only ones which are not helped to ascend again by personal selection, and therefore in their case alone can we form any idea of how the primary constituents behave, when they are subject solely to intra-germinal forces.

The whole determinant system of an id, as I conceive it, is in a state of continual fluctuation upwards and downwards. In most cases the fluctuations will counteract one another, because the passive streams of nutriment soon change, but in many cases the limit from which a return is possible will be passed, and then the determinants concerned will continue to vary in the same direction, till they attain positive or negative selection-value. At this stage personal selection intervenes and sets aside the variation if it is disadvantageous, or favours--that is to say, preserves--it if it is advantageous. Only the determinant of a useless organ is uninfluenced by personal selection, and, as experience shows, it sinks downwards; that is, the organ that corresponds to it degenerates very slowly but uninterruptedly till, after what must obviously be an immense stretch of time, it disappears from the germ-plasm altogether.

Thus we find in the fact of the degeneration of disused parts the proof that not all the fluctuations of a determinant return to equilibrium again, but that, when the movement has attained to a certain strength, it continues in the same direction. We have entire certainty in regard to this as far as the downward progress is concerned, and we must assume it also in regard to ascending variations, as the phenomena of artificial selection certainly justify us in doing. If the Japanese breeders were able to lengthen the tail-feathers of the cock to six feet, it can only have been because the determinants of the tail-feathers in the germ-plasm had already struck out a path of ascending variation, and this movement was taken advantage of by the breeder, who continually selected for reproduction the individuals in which the ascending variation was most marked. For all breeding depends upon the unconscious selection of germinal variations.

Of course these germinal processes cannot be proved mathematically, since we cannot actually see the play of forces of the passive fluctuations and their causes. We cannot say how great these fluctuations are, and how quickly or slowly, how regularly or irregularly they change. Nor do we know how far a determinant must be strengthened by the passive flow of the nutritive stream if it is to be beyond the danger of unfavourable variations, or how far it must be weakened passively before it loses the power of recovering itself by its own strength. It is no more possible to bring forward actual proofs in this case than it was in regard to the selection-value of the initial stages of an adaptation. But if we consider that all heritable variations must have their roots in the germ-plasm, and further, that when personal selection does not intervene, that is to say, in the case of parts which have become useless, a degeneration of the part, and therefore also of its determinant must inevitably take place; then we must conclude that processes such as I have assumed are running their course within the germ-plasm, and we can do this with as much certainty as we were able to infer, from the phenomena of adaptation, the selection-value of their initial stages. The fact of the degeneration of disused parts seems to me to afford irrefutable proof that the fluctuations within the germ-plasm are the real root of all hereditary variation, and the preliminary condition for the occurrence of the Darwin-Wallace factor of selection. Germinal selection supplies the stones out of which personal selection builds her temples and palaces: adaptations. The importance for the theory of the process of degeneration of disused parts cannot be over-estimated, especially when it occurs in sterile animal forms, where we are free from the doubt as to the alleged Lamarckian factor which is apt to confuse our ideas in regard to other cases.

If we regard the variation of the many determinants concerned in the transformation of the female into the sterile worker as having come about through the gradual transformation of the ids into worker-ids, we shall see that the germ-plasm of the sexual ants must contain three kinds of ids, male, female, and worker ids, or if the workers have diverged into soldiers and nest-builders, then four kinds. We understand that the worker-ids arose because their determinants struck out a useful path of variation, whether upward or downward, and that they continued in this path until the highest attainable degree of utility of the parts determined was reached. But in addition to the organs of positive or negative selection-value, there were some which were indifferent as far as the success and especially the functional capacity of the workers was concerned: wings, ovarian tubes, receptaculum seminis, a number of the facets of the eye, perhaps even the whole eye. As to the ovarian tubes it is is possible that their degeneration was an advantage for the workers, in saving energy, and if so selection would favour the degeneration; but how could the presence of eyes diminish the usefulness of the workers to the colony or the minute receptaculum seminis, or even the wings? These parts have therefore degenerated because they were of no further value to the insect. But if selection did not influence the setting aside of these parts because they were neither of advantage nor of disadvantage to the species, then the Darwinian factor of selection is here confronted with a puzzle which it cannot solve alone, but which at once becomes clear when germinal selection is added. For the determinants of organs that have no further value for the organism, must, as we have already explained, embark on a gradual course of retrograde development.

In ants the degeneration has gone so far that there are no wing-rudiments present in any species, as is the case with so many butterflies, flies, and locusts, but in the larvae the imaginable discs of the wings are still laid down. With regard to the ovaries, degeneration has reached different levels in different species of ants, as has been shown by the researches of my former pupil, Elizabeth Bickford. In many species there are twelve ovarian tubes, and they decrease from that number to one; indeed, in one species no ovarian tube at all is present. So much at least is certain from what has been said, that in this case everything depends on the fluctuations of the elements of the germ-plasm. Germinal selection, here as elsewhere, presents the variations of the determinants, and personal selection favours or rejects these, or,--if it be a question of organs which have become useless,--it does not come into play at all, and allows the descending variation free course.

It is obvious that even the problem of coadaptation in sterile animals can thus be satisfactorily explained. If the determinants are oscillating upwards and downwards in continual fluctuation, and varying more pronouncedly now in one direction now in the other, useful variations of every determinant will continually present themselves anew, and may, in the course of generations, be combined with one another in various ways. But there is one character of the determinants that greatly facilitates this complex process of selection, that, after a certain limit has been reached, they go on varying in the same direction. From this it follows that development along a path once struck out may proceed without the continual intervention of personal selection. This factor only operates, so to speak, at the beginning, when it selects the determinants which are varying in the right direction, and again at the end, when it is necessary to put a check upon further variation. In addition to this, enormously long periods have been available for all these adaptations, as the very gradual transition stages between females and workers in many species plainly show, and thus this process of transformation loses the marvellous and mysterious character that seemed at the first glance to invest it, and takes rank, without any straining, among the other processes of selection. It seems to me that, from the facts that sterile animal forms can adapt themselves to new vital functions, their superfluous parts degenerate, and the parts more used adapt themselves in an ascending direction, those less used in a descending direction, we must draw the conclusion that harmonious adaptation here comes about without the cooperation of the Lamarckian principle. This conclusion once established, however, we have no reason to refer the thousands of cases of harmonious adaptation, which occur in exactly the same way among other animals or plants, to a principle, the active intervention of which in the transformation of species is nowhere proved. We do not require it to explain the facts, and therefore we must not assume it.

The fact of coadaptation, which was supposed to furnish the strongest argument against the principle of selection, in reality yields the clearest evidence in favour of it. We must assume it, because no other possibility of explanation is open to us, and because these adaptations actually exist, that is to say, have really taken place. With this conviction I attempted, as far back as 1894, when the idea of germinal selection had not yet occurred to me, to make "harmonious adaptation" (coadaptation) more easily intelligible in some way or other, and so I was led to the idea, which was subsequently expounded in detail by Baldwin, and Lloyd Morgan, and also by Osborn, and Gulick as Organic Selection. It seemed to me that it was not necessary that all the germinal variations required for secondary variations should have occurred simultaneously, since, for instance, in the case of the stag, the bones, muscles, sinews, and nerves would be incited by the increasing heaviness of the antlers to greater activity in the individual life, and so would be strengthened. The antlers can only have increased in size by very slow degrees, so that the muscles and bones may have been able to keep pace with their growth in the individual life, until the requisite germinal variations presented themselves. In this way a disharmony between the increasing weight of the antlers and the parts which support and move them would be avoided, since time would be given for the appropriate germinal variations to occur, and so to set agoing the hereditary variation of the muscles, sinews and bones.[42]

I still regard this idea as correct, but I attribute less importance to "organic selection" than I did at that time, in so far that I do not believe that it alone could effect complex harmonious adaptations. Germinal selection now seems to me to play the chief part in bringing about such adaptations. Something the same is true of the principle I have called Panmixia. As I became more and more convinced, in the course of years, that the Lamarckian principle ought not to be called in to explain the dwindling of disused parts, I believed that this process might be simply explained as due to the cessation of the conservative effect of natural selection. I said to myself that, from the moment in which a part ceases to be of use, natural selection withdraws its hand from it, and then it must inevitably fall from the height of its adaptiveness, because inferior variants would have as good a chance of persisting as better ones, since all grades of fitness of the part in question would be mingled with one another indiscriminately. This is undoubtedly true, as Romanes pointed out ten years before I did, and this mingling of the bad with the good probably does bring about a deterioration of the part concerned. But it cannot account for the steady diminution, which always occurs when a part is in process of becoming rudimentary, and which goes on until it ultimately disappears altogether. The process of dwindling cannot therefore be explained as due to panmixia alone: we can only find a sufficient explanation in germinal selection.

IV. DERIVATIVES OF THE THEORY OF SELECTION

The impetus in all directions given by Darwin through his theory of selection has been an immeasurable one, and its influence is still felt. It falls within the province of the historian of science to enumerate all the ideas which, in the last quarter of the nineteenth century, grew out of Darwin's theories, in the endeavour to penetrate more deeply into the problem of the evolution of the organic world. Within the narrow limits to which this paper is restricted, I cannot attempt to discuss any of these.

V. ARGUMENTS FOR THE REALITY OF THE PROCESSES OF SELECTION

(a) Sexual Selection

Sexual selection goes hand in hand with natural selection. From the very first I have regarded sexual selection as affording an extremely important and interesting corroboration of natural selection, but, singularly enough, it is precisely against this theory that an adverse judgment has been pronounced in so many quarters, and it is only quite recently, and probably in proportion as the wealth of facts in proof of it penetrates into a wider circle, that we seem to be approaching a more general recognition of this side of the problem of adaptations. Thus Darwin's words in his preface to the second edition (1874) of his book, The Descent of Man and Sexual Selection, are being justified: "My conviction as to the operation of natural selection remains unshaken," and further, "If naturalists were to become more familiar with the idea of sexual selection, it would, I think, be accepted to a much greater extent, and already it is fully and favourably accepted by many competent judges." Darwin was able to speak thus because he was already acquainted with an immense mass of facts, which, taken together, yield overwhelming evidence of the validity of the principle of sexual selection.

Natural selection chooses out for reproduction the individuals that are best equipped for the struggle for existence, and it does so at every stage of development; it thus improves the species in all its stages and forms. Sexual selection operates only on individuals that are already capable of reproduction, and does so only in relation to the attainment of reproduction. It arises from the rivalry of one sex, usually the male, for the possession of the other, usually the female. Its influence can therefore only directly affect one sex, in that it equips it better for attaining possession of the other. But the effect may extend indirectly to the female sex, and thus the whole species may be modified, without, however, becoming any more capable of resistance in the struggle for existence, for sexual selection only gives rise to adaptations which are likely to give their possessor the victory over rivals in the struggle for possession of the female, and which are therefore peculiar to the wooing sex: the manifold "secondary sexual characters." The diversity of these characters is so great that I cannot here attempt to give anything approaching a complete treatment of them, but I should like to give a sufficient number of examples to make the principle itself, in its various modes of expression, quite clear.

One of the chief preliminary postulates of sexual selection is the unequal number of individuals in the two sexes, for if every male immediately finds his mate there can be no competition for the possession of the female. Darwin has shown that, for the most part, the inequality between the sexes is due simply to the fact that there are more males than females, and therefore the males must take some pains to secure a mate. But the inequality does not always depend on the numerical preponderance of the males, it is often due to polygamy; for, if one male claims several females, the number of females in proportion to the rest of the males will be reduced. Since it is almost always the males that are the wooers, we must expect to find the occurrence of secondary sexual characters chiefly among them, and to find it especially frequent in polygamous species. And this is actually the case.

If we were to try to guess--without knowing the facts--what means the male animals make use of to overcome their rivals in the struggle for the possession of the female, we might name many kinds of means, but it would be difficult to suggest any which is not actually employed in some animal group of other. I begin with the mere difference in strength, through which the male of many animals is so sharply distinguished from the female, as, for instance, the lion, walrus, "sea-elephant," and others. Among these the males fight violently for the possession of the female, who falls to the victor in the combat. In this simple case no one can doubt the operation of selection, and there is just as little room for doubt as to the selection-value of the initial stages of the variation. Differences in bodily strength are apparent even among human beings, although in their case the struggle for the possession of the female is no longer decided by bodily strength alone.

Combats between male animals are often violent and obstinate, and the employment of the natural weapons of the species in this way has led to perfecting of these, e.g. the tusks of the boar, the antlers of the stag, and the enormous, antler-like jaws of the stag-beetle. Here again it is impossible to doubt that variations in these organs presented themselves, and that these were considerable enough to be decisive in combat, and so to lead to the improvement of the weapon.

Among many animals, however, the females at first withdraw from the males; they are coy, and have to be sought out, and sometimes held by force. This tracking and grasping of the females by the males has given rise to many different characters in the latter, as, for instance, the larger eyes of the male bee, and especially of the males of the Ephemerids (May-flies), some species of which show, in addition to the usual compound eyes, large, so-called turban-eyes, so that the whole head is covered with seeing surfaces. In these species the females are very greatly in the minority (1-100), and it is easy to understand that a keen competition for them must take place, and that, when the insects of both sexes are floating freely in the air, an unusually wide range of vision will carry with it a decided advantage. Here again the actual adaptations are in accordance with the preliminary postulates of the theory. We do not know the stages through which the eye has passed to its present perfected state, but, since the number of simple eyes (facets) has become very much greater in the male than in the female, we may assume that their increase is due to a gradual duplication of the determinants of the ommatidium in the germ-plasm, as I have already indicated in regard to sense-organs in general. In this case, again, the selection-value of the initial stages hardly admits of doubt; better vision directly secures reproduction.

In many cases the organ of smell shows a similar improvement. Many lower Crustaceans (Daphnidae) have better developed organs of smell in the male sex. The difference is often slight and amounts only to one or two olfactory filaments, but certain species show a difference of nearly a hundred of these filaments (Leptodora). The same thing occurs among insects.

We must briefly consider the clasping or grasping organs which have developed in the males among many lower Crustaceans, but here natural selection plays its part along with sexual selection, for the union of the sexes is an indispensable condition for the maintenance of the species, and as Darwin himself pointed out, in many cases the two forms of selection merge into each other. This fact has always seemed to me to be a proof of natural selection, for, in regard to sexual selection, it is quite obvious that the victory of the best-equipped could have brought about the improvement only of the organs concerned, the factors in the struggle, such as the eye and the olfactory organ.

We come now to the excitants; that is, to the group of sexual characters whose origin through processes of selection has been most frequently called in question. We may cite the love-calls produced by many male insects, such as crickets and cicadas. These could only have arisen in animal groups in which the female did not rapidly flee from the male, but was inclined to accept his wooing from the first. Thus, notes like the chirping of the male cricket serve to entice the females. At first they were merely the signal which showed the presence of a male in the neighbourhood, and the female was gradually enticed nearer and nearer by the continued chirping. The male that could make himself heard to the greatest distance would obtain the largest following, and would transmit the beginnings, and, later, the improvement of his voice to the greatest number of descendants. But sexual excitement in the female became associated with the hearing of the love-call, and then the sound-producing organ of the male began to improve, until it attained to the emission of the long-drawn-out soft notes of the mole-cricket or the maenad-like cry of the cicadas. I cannot here follow the process of development in detail, but will call attention to the fact that the original purpose of the voice, the announcing of the male's presence, became subsidiary, and the exciting of the female became the chief goal to be aimed at. The loudest singers awakened the strongest excitement, and the improvement resulted as a matter of course. I conceive of the origin of bird-song in a somewhat similar manner, first as a means of enticing, then of exciting the female.

One more kind of secondary sexual character must here be mentioned: the odour which emanates from so many animals at the breeding season. It is possible that this odour also served at first merely to give notice of the presence of individuals of the other sex, but it soon became an excitant, and as the individuals which caused the greatest degree of excitement were preferred, it reached as high a pitch of perfection as was possible to it. I shall confine myself here to the comparatively recently discovered fragrance of butterflies. Since Fritz Muller found out that certain Brazilian butterflies gave off fragrance "like a flower," we have become acquainted with many such cases, and we now know that in all lands, not only many diurnal Lepidoptera but nocturnal ones also give off a delicate odour, which is agreeable even to man. The ethereal oil to which this fragrance is due is secreted by the skin-cells, usually of the wing, as I showed soon after the discovery of the scent-scales. This is the case in the males; the females have no special scent-scales recognisable as such by their form, but they must, nevertheless, give off an extremely delicate fragrance, although our imperfect organ of smell cannot perceive it, for the males become aware of the presence of a female, even at night, from a long distance off, and gather round her. We may therefore conclude, that both sexes have long given forth a very delicate perfume, which announced their presence to others of the same species, and that in many species (not in all) these small beginnings become, in the males, particularly strong scent-scales of characteristic form (lute, brush, or lyre-shaped). At first these scales were scattered over the surface of the wing, but gradually they concentrated themselves, and formed broad, velvety bands, or strong, prominent brushes, and they attained their highest pitch of evolution when they became enclosed within pits or folds of the skin, which could be opened to let the delicious fragrance stream forth suddenly towards the female. Thus in this case also we see that characters, the original use of which was to bring the sexes together, and so to maintain the species, have been evolved in the males into means for exciting the female. And we can hardly doubt, that the females are most readily enticed to yield to the butterfly that sends out the strongest fragrance,--that is to say, that excites them to the highest degree. It is a pity that our organs of smell are not fine enough to examine the fragrance of male Lepidoptera in general, and to compare it with other perfumes which attract these insects.[43] As far as we can perceive them they resemble the fragrance of flowers, but there are Lepidoptera whose scent suggests musk. A smell of musk is also given off by several plants: it is a sexual excitant in the musk-deer, the musk-sheep, and the crocodile.

As far as we know, then, it is perfumes similar to those of flowers that the male Lepidoptera give off in order to entice their mates and this is a further indication that animals, like plants, can to a large extent meet the claims made upon them by life, and produce the adaptations which are most purposive,--a further proof, too, of my proposition that the useful variations, so to speak, are always there. The flowers developed the perfumes which entice their visitors, and the male Lepidoptera developed the perfumes which entice and excite their mates.

There are many pretty little problems to be solved in this connection, for there are insects, such as some flies, that are attracted by smells which are unpleasant to us, like those from decaying flesh and carrion. But there are also certain flowers, some orchids for instance, which give forth no very agreeable odour, but one which is to us repulsive and disgusting; and we should therefore expect that the males of such insects would give off a smell unpleasant to us, but there is no case known to me in which this has been demonstrated.

In cases such as we have discussed, it is obvious that there is no possible explanation except through selection. This brings us to the last kind of secondary sexual characters, and the one in regard to which doubt has been most frequently expressed,--decorative colours and decorative forms, the brilliant plumage of the male pheasant, the humming-birds, and the bird of Paradise, as well as the bright colours of many species of butterfly, from the beautiful blue of our little Lycaenidae to the magnificent azure of the large Morphinae of Brazil. In a great many cases, though not by any means in all, the male butterflies are "more beautiful" than the females, and in the Tropics in particular they shine and glow in the most superb colours. I really see no reason why we should doubt the power of sexual selection, and I myself stand wholly on Darwin's side. Even though we certainly cannot assume that the females exercise a conscious choice of the "handsomest" mate, and deliberate like the judges in a court of justice over the perfections of their wooers, we have no reason to doubt that distinctive forms (decorative feathers), and colours have a particularly exciting effect upon the female, just as certain odours have among animals of so many different groups, including the butterflies. The doubts which existed for a considerable time, as a result of fallacious experiments, as to whether the colours of flowers really had any influence in attracting butterflies have now been set at rest through a series of more careful investigations; we now know that the colours of flowers are there on account of the butterflies, as Sprengel first showed, and that the blossoms of Phanerogams are selected in relation to them, as Darwin pointed out.

Certainly it is not possible to bring forward any convincing proof of the origin of decorative colours through sexual selection, but there are many weighty arguments in favour of it, and these form a body of presumptive evidence so strong that it almost amounts to certainty.

In the first place, there is the analogy with other secondary sexual characters. If the song of birds and the chirping of the cricket have been evolved through sexual selection, if the penetrating odours of male animals,--the crocodile, the musk-deer, the beaver, the carnivores, and, finally, the flower-like fragrances of the butterflies have been evolved to their present pitch in this way, why should decorative colours have arisen in some other way? Why should the eye be less sensitive to specifically male colours and other visible signs enticing to the female, than the olfactory sense to specifically male odours, or the sense of hearing to specifically male sounds?

Moreover, the decorative feathers of birds are almost always spread out and displayed before the female during courtship. I have elsewhere[44] pointed out that decorative colouring and sweet-scentedness may replace one another in Lepidoptera as well as in flowers, for just as some modestly coloured flowers (mignonette and violet) have often a strong perfume, while strikingly coloured ones are sometimes quite devoid of fragrance, so we find that the most beautiful and gaily-coloured of our native Lepidoptera, the species of Vanessa, have no scent-scales, while these are often markedly developed in grey nocturnal Lepidoptera. Both attractions may, however, be combined in butterflies, just as in flowers. Of course, we cannot explain why both means of attraction should exist in one genus, and only one of them in another, since we do not know the minutest details of the conditions of life of the genera concerned. But from the sporadic distribution of scent-scales in Lepidoptera, and from their occurrence or absence in nearly related species, we may conclude that fragrance is a relatively modern acquirement, more recent than brilliant colouring.

One thing in particular that stamps decorative colouring as a product of selection is its gradual intensification by the addition of new spots, which we can quite well observe, because in many cases the colours have been first acquired by the males, and later transmitted to the females by inheritance. The scent-scales are never thus transmitted, probably for the same reason that the decorative colours of many birds are often not transmitted to the females: because with these they would be exposed to too great elimination by enemies. Wallace was the first to point out that in species with concealed nests the beautiful feathers of the male occurred in the female also, as in the parrots, for instance, but this is not the case in species which brood on an exposed nest. In the parrots one can often observe that the general brilliant colouring of the male is found in the female, but that certain spots of colour are absent, and these have probably been acquired comparatively recently by the male and have not yet been transmitted to the female.

Isolation of the group of individuals which is in process of varying is undoubtedly of great value in sexual selection, for even a solitary conspicuous variation will become dominant much sooner in a small isolated colony, than among a large number of members of a species.

Any one who agrees with me in deriving variations from germinal selection will regard that process as an essential aid towards explaining the selection of distinctive courtship-characters, such as coloured spots, decorative feathers, horny outgrowths in birds and reptiles, combs, feather-tufts, and the like, since the beginnings of these would be presented with relative frequency in the struggle between the determinants within the germ-plasm. The process of transmission of decorative feathers to the female results, as Darwin pointed out and illustrated by interesting examples, in the colour-transformation of a whole species, and this process, as the phyletically older colouring of young birds shows, must, in the course of thousands of years, have repeated itself several times in a line of descent.

If we survey the wealth of phenomena presented to us by secondary sexual characters, we can hardly fail to be convinced of the truth of the principle of sexual selection. And certainly no one who has accepted natural selection should reject sexual selection, for, not only do the two processes rest upon the same basis, but they merge into one another, so that it is often impossible to say how much of a particular character depends on one and how much on the other form of selection.

(b) Natural Selection

An actual proof of the theory of sexual selection is out of the question, if only because we cannot tell when a variation attains to selection-value. It is certain that a delicate sense of smell is of value to the male moth in his search for the female, but whether the possession of one additional olfactory hair, or of ten, or of twenty additional hairs leads to the success of its possessor we are unable to tell. And we are groping even more in the dark when we discuss the excitement caused in the female by agreeable perfumes, or by striking and beautiful colours. That these do make an impression is beyond doubt; but we can only assume that slight intensifications of them give any advantage, and we must assume this since otherwise secondary sexual characters remain inexplicable.

The same thing is true in regard to natural selection. It is not possible to bring forward any actual proof of the selection-value of the initial stages, and the stages in the increase of variations, as has been already shown. But the selection-value of a finished adaptation can in many cases be statistically determined. Cesnola and Poulton have made valuable experiments in this direction. The former attached forty-five individuals of the green, and sixty-five of the brown variety of the praying mantis (Mantis religiosa), by a silk thread to plants, and watched them for seven days. The insects which were on a surface of a colour Similar to their own remained uneaten, while twenty-five green insects on brown parts of plants had all disappeared in eleven days.

The experiments of Poulton and Sanders[45] were made with 600 pupae of Vanessa urticae, the "tortoise-shell butterfly." The pupae were artificially attached to nettles, tree-trunks, fences, walls, and to the ground, some at Oxford, some at St. Helens in the Isle of Wight. In the course of a month 93% of the pupae at Oxford were killed, chiefly by small birds, while at St. Helens 68% perished. The experiments showed very clearly that the colour and character of the surface on which the pupa rests--and thus its own conspicuousness--are of the greatest importance. At Oxford only the four pupae which were fastened to nettles emerged; all the rest--on bark, stones and the like--perished. At St. Helens the elimination was as follows: on fences where the pupae were conspicuous, 92%; on bark, 66%; on walls, 54%; and among nettles, 57%. These interesting experiments confirm our views as to protective coloration, and show further, that the ratio of elimination in the species is a very high one, and that therefore selection must be very keen.

We may say that the process of selection follows as a logical necessity from the fulfilment of the three preliminary postulates of the theory: variability, heredity, and the struggle for existence, with its enormous ratio of elimination in all species. To this we must add a fourth factor, the intensification of variations which Darwin established as a fact, and which we are now able to account for theoretically on the basis of germinal selection. It may be objected that there is considerable uncertainty about this logical proof, because of our inability to demonstrate the selection-value of the initial stages and the individual stages of increase. We have therefore to fall back on presumptive evidence. This is to be found in the interpretative value of the theory. Let us consider this point in greater detail.

In the first place it is necessary to emphasize what is often overlooked, namely, that the theory not only explains the transformations of species, it also explains their remaining the same; in addition to the principle of varying, it contains within itself that of persisting. It is part of the essence of selection, that it not only causes a part to vary till it has reached its highest pitch of adaptation, but that it maintains it at this pitch. This conserving influence of natural selection is of great importance, and was early recognised by Darwin; it follows naturally from the principle of the survival of the fittest.

We understand from this how it is that a species which has become fully adapted to certain conditions of life ceases to vary, but remains "constant," as long as the conditions of life for it remain unchanged, whether this be for thousands of years, or for whole geological epochs. But the most convincing proof of the power of the principle of selection lies in the innumerable multitude of phenomena which cannot be explained in any other way. To this category belong all structures which are only passively of advantage to the organism, because none of these can have arisen by the alleged Lamarckian principle. These have been so often discussed that we need do no more than indicate them here. Until quite recently the sympathetic coloration of animals--for instance, the whiteness of Arctic animals--was referred, at least in part, to the direct influence of external factors, but the facts can best be explained by referring them to the processes of selection, for then it is unnecessary to make the gratuitous assumption that many species are sensitive to the stimulus of cold and that others are not. The great majority of Arctic land-animals, mammals and birds, are white, and this proves that they were all able to present the variation which was most useful for them. The sable is brown, but it lives in trees, where the brown colouring protects and conceals it more effectively. The musk-sheep (Ovibos moschatus) is also brown, and contrasts sharply with the ice and snow, but it is protected from beasts of prey by its gregarious habit, and therefore it is of advantage to be visible from as great a distance as possible. That so many species have been able to give rise to white varieties does not depend on a special sensitiveness of the skin to the influence of cold, but to the fact that Mammals and Birds have a general tendency to vary towards white. Even with us, many birds-- starlings, blackbirds, swallows, etc.--occasionally produce white individuals, but the white variety does not persist, because it readily falls a victim to the carnivores. This is true of white fawns, foxes, deer, etc. The whiteness, therefore, arises from internal causes, and only persists when it is useful. A great many animals living in a green environment have become clothed in green, especially insects, caterpillars, and Mantidae, both persecuted and persecutors.

That it is not the direct effect of the environment which calls forth the green colour is shown by the many kinds of caterpillar which rest on leaves and feed on them, but are nevertheless brown. These feed by night and betake themselves through the day to the trunk of the tree, and hide in the furrows of the bark. We cannot, however, conclude from this that they were unable to vary towards green, for there are Arctic animals which are white only in winter and brown in summer (Alpine hare, and the ptarmigan of the Alps), and there are also green leaf-insects which remain green only while they are young and difficult to see on the leaf, but which become brown again in the last stage of larval life, when they have outgrown the leaf. They then conceal themselves by day, sometimes only among withered leaves on the ground, sometimes in the earth itself. It is interesting that in one genus, Chaerocampa, one species is brown in the last stage of larval life, another becomes brown earlier, and in many species the last stage is not wholly brown, a part remaining green. Whether this is a case of a double adaptation, or whether the green is being gradually crowded out by the brown, the fact remains that the same species, even the same individual, can exhibit both variations. The case is the same with many of the leaf-like Orthoptera, as, for instance, the praying mantis (Mantis religiosa) which we have already mentioned.

But the best proofs are furnished by those of ten-cited cases in which the insect bears a deceptive resemblance to another object. We now know many such cases, such as the numerous imitations of green or withered leaves, which are brought about in the most diverse ways, sometimes by mere variations in the form of the insect and in its colour, sometimes by an elaborate marking, like that which occurs in the Indian leaf-butterflies, Kallima inachis. In the single butterfly-genus Anaea, in the woods of South America, there are about a hundred species which are all gaily coloured on the upper surface, and on the reverse side exhibit the most delicate imitation of the colouring and pattern of a leaf, generally without any indication of the leaf-ribs, but extremely deceptive nevertheless. Anyone who has seen only one such butterfly may doubt whether many of the insignificant details of the marking can really be of advantage to the insect. Such details are for instance the apparent holes and splits in the apparently dry or half-rotten leaf, which are usually due to the fact that the scales are absent on a circular or oval patch so that the colourless wing-membrane lies bare, and one can look through the spot as through a window. Whether the bird which is seeking or pursuing the butterflies takes these holes for dewdrops, or for the work of a devouring insect, does not affect the question; the mirror-like spot undoubtedly increases the general deceptiveness, for the same thing occurs in many leaf-butterflies, though not in all, and in some cases it is replaced in quite a peculiar manner. In one species of Anaea (A. divina), the resting butterfly looks exactly like a leaf out of the outer edge of which a large semi-circular piece has been eaten, possibly by a caterpillar; but if we look more closely it is obvious that there is no part of the wing absent, and that the semi-circular piece is of a clear, pale yellow colour, while the rest of the wing is of a strongly contrasted dark brown.

But the deceptive resemblance may be caused in quite a different manner. I have often speculated as to what advantage the brilliant white C could give to the otherwise dusky-coloured "Comma butterfly" (Grapta C. album). Poulton's recent observations[46] have shown that this represents the imitation of a crack such as is often seen in dry leaves, and is very conspicuous because the light shines through it.

The utility obviously lies in presenting to the bird the very familiar picture of a broken leaf with a clear shining slit, and we may conclude, from the imitation of such small details, that the birds are very sharp observers and that the smallest deviation from the usual arrests their attention and incites them to closer investigation. It is obvious that such detailed--we might almost say such subtle--deceptive resemblances could only have come about in the course of long ages through the acquirement from time to time of something new which heightened the already existing resemblance.

In face of facts like these there can be no question of chance and no one has succeeded so far in finding any other explanation to replace that by selection. For the rest, the apparent leaves are by no means perfect copies of a leaf; many of them only represent the torn or broken piece, or the half or two-thirds of a leaf, but then the leaves themselves frequently do not present themselves to the eye as a whole, but partially concealed among other leaves. Even those butterflies which, like the species of Kallima and Anaea, represent the whole of a leaf with stalk, ribs, apex, and the whole breadth, are not actual copies which would satisfy a botanist; there is often much wanting. In Kallima the lateral ribs of the leaf are never all included in the markings; there are only two or three on the left side and at more four or five on the right, and in many individuals these are rather obscure, while in others they are comparatively distinct. This furnishes us with fresh evidence in favour of their origin through processes of selection, for a botanically perfect picture could not arise in this way; there could only be a fixing of such details as heightened the deceptive resemblance.

Our postulate of origin through selection also enables us to understand why the leaf-imitation is on the lower surface of the wing in the diurnal Lepidoptera, and on the upper surface in the nocturnal forms, corresponding to the attitude of the wings in the resting position of the two groups.

The strongest of all proofs of the theory, however, is afforded by cases of true "mimicry," those adaptations discovered by Bates in 1861, consisting in the imitation of one species by another, which becomes more and more like its model. The model is always a species that enjoys some special protection from enemies, whether because it is unpleasant to taste, or because it is in some way dangerous.

It is chiefly among insects and especially among butterflies that we find the greatest number of such cases. Several of these have been minutely studied and every detail has been investigated so that it is difficult to understand how there can still be disbelief in regard to them. If the many and exact observations which have been carefully collected and critically discussed for instance by Poulton[47] were thoroughly studied the arguments which are still frequently urged against mimicry would be found untenable; we can hardly hope to find more convincing proof of the actuality of the processes of selection than these cases put into our hands. The preliminary postulates of the theory of mimicry have been disputed, for instance, that diurnal butterflies are persecuted and eaten by birds, but observations specially directed towards this point in India, Africa, America and Europe have placed it beyond all doubt. If it were necessary I could myself furnish an account of my own observations on this point.

In the same way it has been established by experiment and observation in the field that in all the great regions of distribution there are butterflies which are rejected by birds and lizards, their chief enemies, on account of their unpleasant smell or taste. These butterflies are usually gaily and conspicuously coloured and thus--as Wallace first interpreted it--are furnished with an easily recognisable sign: a sign of unpalatableness or warning colours. If they were not thus recognisable easily and from a distance, they would frequently be pecked at by birds, and then rejected because of their unpleasant taste; but as it is, the insect-eaters recognise them at once as unpalatable booty and ignore them. Such immune[48] species, wherever they occur, are imitated by other palatable species, which thus acquire a certain degree of protection.

It is true that this explanation of the bright, conspicuous colours is only a hypothesis, but its foundations--unpalatableness, and the liability of other butterflies to be eaten,--are certain, and its consequences--the existence of mimetic palatable forms--conform it in the most convincing manner. Of the many cases now known I select one, which is especially remarkable, and which has been thoroughly investigated, Papilla dardanus (merope), a large, beautiful, diurnal butterfly which ranges from Abyssinia throughout the whole of Africa to the south coast of Cape Colony.

The males of this form are everywhere almost the same in colour and in form of wings, save for a few variations in the sparse black markings on the pale yellow ground. But the females occur in several quite different forms and colourings, and one of these only, the Abyssinian form, is like the male, while the other three or four are mimetic, that is to say, they copy a butterfly of quite a different family the Danaids, which are among the immune forms. In each region the females have thus copied two or three different immune species. There is much that is interesting to be said in regard to these species, but it would be out of keeping with the general tenor of this paper to give details of this very complicated case of polymorphism in P. Dardanus. Anyone who is interested in the matter will find a full and exact statement of the case in as far as we know it, in Poulton's Essays on Evolution (pp. 373-375[49]). I need only add that three different mimetic female forms have been reared from the eggs of a single female in South Africa. The resemblance of the forms to their immune models goes so far that even the details of the local forms of the models are copied by the mimetic species.

It remains to be said that in Madagascar a butterfly,

Papilio meriones, occurs, of which both sexes are very similar in form and markings to the non-mimetic male of P. dardanus, so that it probably represents the ancestor of this latter species.

In face of such facts as these every attempt at another explanation must fail. Similarly all the other details of the case fulfil the preliminary postulates of selection, and leave no room for any other interpretation. That the males do not take on the protective colouring is easily explained, because they are in general more numerous, and the females are more important for the preservation of the species, and must also live longer in order to deposit their eggs. We find the same state of things in many other species, and in one case (Elymnias undularis) in which the male is also mimetically coloured, it copies quite a differently coloured immune species from the model followed by the female. This is quite intelligible when we consider that if there were too many false immune types, the birds would soon discover that there were palatable individuals among those with unpalatable warning colours. Hence the imitation of different immune species by Papilio dardanus!

I regret that lack of space prevents my bringing forward more examples of mimicry and discussing them fully. But from the case of Papilio dardanus alone there is much to be learnt which is of the highest importance for our understanding of transformations. It shows us chiefly what I once called, somewhat strongly perhaps, the omnipotence of natural selection in answer to an opponent who had spoken of its "inadequacy." We here see that one and the same species is capable of producing four or five different patterns of colouring and marking; thus the colouring and marking are not, as has often been supposed, a necessary outcome of the specific nature of the species, but a true adaptation, which cannot arise as a direct effect of climatic conditions, but solely through what I may call the sorting out of the variations produced by the species, according to their utility. That caterpillars may be either green or brown is already something more than could have been expected according to the old conception of species, but that one and the same butterfly should be now pale yellow, with black; now red with black and pure white; now deep black with large, pure white spots; and again black with a large ocheous-yellow spot, and many small white and yellow spots; that in one sub-species it may be tailed like the ancestral form, and in another tailless like its Danaid model,--all this shows a far-reaching capacity for variation and adaptation that we could never have expected if we did not see the facts before us. How it is possible that the primary colour-variations should thus be intensified and combined remains a puzzle even now; we are reminded of the modern three-colour printing,--perhaps similar combinations of the primary colours take place in this case; in any case the direction of these primary variations is determined by the artist whom we know as natural selection, for there is no other conceivable way in which the model could affect the butterfly that is becoming more and more like it. The same climate surrounds all four forms of female; they are subject to the same conditions of nutrition. Moreover, Papilio dardanus is by no means the only species of butterfly which exhibits different kinds of colour-pattern on its wings. Many species of the Asiatic genus Elymnias have on the upper surface a very good imitation of an immune Euploeine (Danainae), often with a steel-blue ground-colour, while the under surface is well concealed when the butterfly is at rest,--thus there are two kinds of protective coloration each with a different meaning! The same thing may be observed in many non-mimetic butterflies, for instance in all our species of Vanessa, in which the under side shows a grey-brown or brownish-black protective coloration, but we do not yet know with certainty what may be the biological significance of the gaily coloured upper surface.

In general it may be said that mimetic butterflies are comparatively rare species, but there are exceptions, for instance Limenitis archippus in North America, of which the immune model (Danaida plexippus) also occurs in enormous numbers.

In another mimicry-category the imitators are often more numerous than the models, namely in the case of the imitation of dangerous insects by harmless species. Bees and wasps are dreaded for their sting, and they are copied by harmless flies of the genera Eristalis and Syrphus, and these mimics often occur in swarms about flowering plants without damage to themselves or to their models; they are feared and are therefore left unmolested.

In regard also to the faithfulness of the copy the facts are quite in harmony with the theory, according to which the resemblance must have arisen and increased by degrees. We can recognise this in many cases, for even now the mimetic species show very varying degrees of resemblance to their immune model. If we compare, for instance, the many different imitators of Danaida chrysippus we find that, with their brownish-yellow ground-colour, and the position and size, and more or less sharp limitation of their clear marginal spots, they have reached very different degrees of nearness to their model. Or compare the female of Elymnias undularis with its model Danaida genutia; there is a general resemblance, but the marking of the Danaida is very roughly imitated in Elymnias.

Another fact that bears out the theory of mimicry is, that even when the resemblance in colour-pattern is very great, the wing-venation, which is so constant, and so important in determining the systematic position of butterflies, is never affected by the variation. The pursuers of the butterfly have no time to trouble about entomological intricacies.

I must not pass over a discovery of Poulton's which is of great theoretical importance--that mimetic butterflies may reach the same effect by very different means.[50] Thus the glass-like transparency of the wing of a certain Ithomiine (Methona) and its Pierine mimic (Dismorphia orise) depends on a diminution in the size of the scales; in the Danaine genus Itune it is due to the fewness of the scales and in a third imitator, a moth (Castnia linus var. heliconoides) the glass-like appearance of the wing is due neither to diminution nor to absence of scales, but to their absolute colourlessness and transparency, and to the fact that they stand upright. In another moth mimic (Anthomyza) the arrangement of the transparent scales is normal. Thus it is not some unknown external influence that has brought about the transparency of the wing in these five forms, as has sometimes been supposed. Nor is it a hypothetical internal evolutionary tendency, for all three vary in a different manner. The cause of this agreement can only lie in selection, which preserves and intensifies in each species the favourable variations that present themselves. The great faithfulness of the copy is astonishing in these cases, for it is not the whole wing which is transparent; certain markings are black in colour, and these contrast sharply with the glass-like ground. It is obvious that the pursuers of these butterflies must be very sharp-sighted, for otherwise the agreement between the species could never have been pushed so far. The less the enemies see and observe, the more defective must the imitation be, and if they had been blind, no visible resemblance between the species which required protection could ever have arisen.

A seemingly irreconcilable contradiction to the mimicry theory is presented in the following cases, which were known to Bates, who, however, never succeeded in bringing them into line with the principle of mimicry.

In South America there are, as we have already said, many mimics of the immune Ithomiinae (or as Bates called them Heliconidae). Among these there occur not merely species which are edible, and thus require the protection of a disguise, but others which are rejected on account of their unpalatableness. How could the Ithomiine dress have developed in their case, and of what use is it, since the species would in any case be immune? In Eastern Brazil, for instance, there are four butterflies, which bear a most confusing resemblance to one another in colour, marking, and form of wing, and all four are unpalatable to birds. They belong to four different genera and three sub-families, and we have to inquire: Whence came this resemblance and what end does it serve? For a long time no satisfactory answer could be found, but Fritz M 黙 ler,[51] seventeen years after Bates, offered a solution to the riddle, when he pointed out that young birds could not have an instinctive knowledge of the unpalatableness of the Ithomiines, but must learn by experience which species were edible and which inedible. Thus each young bird must have tasted at least one individual of each inedible species and discovered its unpalatability, before it learnt to avoid, and thus to spare the species. But if the four species resemble each other very closely the bird will regard them all as of the same kind, and avoid them all. Thus there developed a process of selection which resulted in the survival of the Ithomiine-like individuals, and in so great an increase of resemblance between the four species, that they are difficult to distinguish one from another even in a collection. The advantage for the four species, living side by side as they do e.g. in Bahia, lies in the fact that only one individual from the mimicry-ring ("inedible association") need be tasted by a young bird, instead of at least four individuals, as would otherwise be the case. As the number of young birds is great, this makes a considerable difference in the ratio of elimination. The four Brazilian species are Lycorea halia (Danainae), Heliconius narcaea (eucrate) (Heliconinae), Melinaea ethra, and Mechanitis lysimnia (Ithomiinae).

These interesting mimicry-rings (trusts), which have much significance for the theory, have been the subject of numerous and careful investigations, and at least their essential features are now fully established. Muller took for granted, without making any investigations, that young birds only learn by experience to distinguish between different kinds of victims. But Lloyd Morgan's[52] experiments with young birds proved that this is really the case, and at the same time furnished an additional argument against the Lamarckian principle.

In addition to the mimicry-rings first observed in South America, others have been described from Tropical India by Moore, and by Poulton and Dixey from Africa, and we may expect to learn many more interesting facts in this connection. Here again the preliminary postulates of the theory are satisfied. And how much more that would lead to the same conclusion might be added!

As in the case of mimicry many species have come to resemble one another through processes of selection, so we know whole classes of phenomena in which plants and animals have become adapted to one another, and have thus been modified to a considerable degree. I refer particularly to the relation between flowers and insects. Darwin has shown that the originally inconspicuous blossoms of the phanerogams were transformed into flowers through the visits of insects, and that, conversely, several large orders of insects have been gradually modified by their association with flowers, especially as regards the parts of their body actively concerned. Bees and butterflies in particular have become what they are through their relation to flowers. In this case again all that is apparently contradictory to the theory can, on closer investigation, be beautifully interpreted in corroboration of it. Selection can give rise only to what is of use to the organism actually concerned, never to what is of use to some other organism, and we must therefore expect to find that in flowers only characters of use to themselves have arisen, never characters which are of use to insects only, and conversely that in the insects characters useful to them and not merely to the plants would have originated. For a long time it seemed as if an exception to this rule existed in the case of the fertilisation of the yucca blossoms by a little moth, Pronuba yuccasella. This little moth has a sickle-shaped appendage to its mouth-parts which occurs hi no other Lepidopteron, and which is used for pushing the yellow pollen into the opening of the pistil, thus fertilising the flower. Thus it appears as if a new structure, which is useful only to the plant, has arisen in the insect. But the difficulty is solved as soon as we learn that the moth lays its eggs in the fruit-buds of the Yucca, and that the larvae, when they emerge, feed on the developing seeds. In effecting the fertilisation of the flower the moth is at the same time making provision for its own offspring, since it is only after fertilisation that the seeds begin to develop. There is thus nothing to prevent our referring this structural adaptation in Pronuba yuccasella to processes of selection, which have gradually transformed the maxillary palps of the female into the sickle-shaped instrument for collecting the pollen, and which have at the same time developed in the insect the instinct to press the pollen into the pistil.

In this domain, then, the theory of selection finds nothing but corroboration, and it would be impossible to substitute for it any other explanation, which now that the facts are so well known, could be regarded as a serious rival to it. That selection is a factor, and a very powerful factor in the evolution of organisms, can no longer be doubted. Even although we cannot bring forward formal proofs of it in detail, cannot calculate definitely the size of the variations which present themselves, and their selection-value, cannot, in short, reduce the whole process to a mathematical formula, yet we must assume selection, because it is the only possible explanation applicable to whole classes of phenomena, and because, on the other hand, it is made up of factors which we know can be proved actually to exist, and which, if they exist, must of logical necessity cooperate in the manner required by the theory. We must accept it because the phenomena of evolution and adaptation must have a natural basis, and because it is the only possible explanation of them.[53]

Many people are willing to admit that selection explains adaptations, but they maintain that only a part of the phenomena are thus explained, because everything does not depend upon adaptation. They regard adaptation as, so to speak, a special effort on the part of Nature, which she keeps in readiness to meet particularly difficult claims of the external world on organisms. But if we look at the matter more carefully we shall find that adaptations are by no means exceptional, but that they are present everywhere in such enormous numbers, that it would be difficult in regard to any structure whatever, to prove that adaptation had not played a part in its evolution.

How often has the senseless objection been urged against selection that it can create nothing, it can only reject. It is true that it cannot create either the living substance or the variations of it; both must be given. But in rejecting one thing it preserves another, intensifies it, combines it, and in this way creates what is new. Everything in organisms depends on adaptation; that is to say, everything must be admitted through the narrow door of selection, otherwise it can take no part in the building up of the whole. But, it is asked, what of the direct effect of external conditions, temperature, nutrition, climate and the like? Undoubtedly these can give rise to variations, but they too must pass through the door of selection, and if they cannot do this they are rejected, eliminated from the constitution of the species.

It may, perhaps, be objected that such external influences are often of a compelling power, and that every animal must submit to them, and that thus selection has no choice and can neither select nor reject. There may be such cases; let us assume for instance that the effect of the cold of the Arctic regions was to make all the mammals become black; the result would be that they would all be eliminated by selection, and that no mammals would be able to live there at all. But in most cases a certain percentage of animals resists these strong influences, and thus selection secures a foothold on which to work, eliminating the unfavourable variation, and establishing a useful colouring, consistent with what is required for the maintenance of the species.

Everything depends upon adaptation! We have spoken much of adaptation in colouring, in connection with the examples brought into prominence by Darwin, because these are conspicuous, easily verified, and at the same time convincing for the theory of selection. But is it only desert and polar animals whose colouring is determined through adaptation? Or the leaf-butterflies, and the mimetic species, or the terrifying markings, and "warning-colours" and a thousand other kinds of sympathetic colouring? It is, indeed, never the colouring alone which makes up the adaptation; the structure of the animal plays a part, often a very essential part, in the protective disguise, and thus many variations may cooperate towards one common end. And it is to be noted that it is by no means only external parts that are changed; internal parts are always modified at the same time--for instance, the delicate elements of the nervous system on which depend the instinct of the insect to hold its wings, when at rest, in a perfectly definite position, which, in the leaf-butterfly, has the effect of bringing the two pieces on which the marking occurs on the anterior and posterior wing into the same direction, and thus displaying as a whole the fine curve of the midrib on the seeming leaf. But the wing-holding instinct is not regulated in the same way in all leaf-butterflies; even our indigenous species of Vanessa, with their protective ground-colouring, have quite a distinctive way of holding their wings so that the greater part of the anterior wing is covered by the posterior when the butterfly is at rest. But the protective colouring appears on the posterior wing and on the tip of the anterior, to precisely the distance to which it is left uncovered. This occurs, as Standfuss has shown, in different degrees in our two most nearly allied species, the uncovered portion being smaller in V. urticae than in V. polychloros. In this case, as in most leaf-butterflies, the holding of the wing was probably the primary character; only after that was thoroughly established did the protective marking develop. In any case, the instinctive manner of holding the wings is associated with the protective colouring, and must remain as it is if the latter is to be effective. How greatly instincts may change, that is to say, may be adapted, is shown by the case of the Noctuid "shark" moth, Xylina vetusta. This form bears a most deceptive resemblance to a piece of rotten wood, and the appearance is greatly increased by the modification of the innate impulse to flight common to so many animals, which has here been transformed into an almost contrary instinct. This moth does not fly away from danger, but "feigns death," that is, it draws antennae, legs and wings close to the body, and remains perfectly motionless. It may be touched, picked up, and thrown down again, and still it does not move. This remarkable instinct must surely have developed simultaneously with the wood-colouring; at all events, both cooperating variations are now present, and prove that both the external and the most minute internal structure have undergone a process of adaptation.

The case is the same with all structural variations of animal parts, which are not absolutely insignificant. When the insects acquired wings they must also have acquired the mechanism with which to move them--the musculature, and the nervous apparatus necessary for its automatic regulation. All instincts depend upon compound reflex mechanisms and are just as indispensable as the parts they have to set in motion, and all may have arisen through processes of selection if the reasons which I have elsewhere given for this view are correct.[54]

Thus there is no lack of adaptations within the organism, and particularly in its most important and complicated parts, so that we may say that there is no actively functional organ that has not undergone a process of adaptation relative to its function and the requirements of the organism. Not only is every gland structurally adapted, down to the very minutest histological details, to its function, but the function is equally minutely adapted to the needs of the body. Every cell in the mucous lining of the intestine is exactly regulated in its relation to the different nutritive substances, and behaves in quite a different way towards the fats, and towards nitrogenous substances, or peptones.

I have elsewhere called attention to the many adaptations of the whale to the surrounding medium, and have pointed out--what has long been known, but is not universally admitted, even now--that in it a great number of important organs have been transformed in adaptation to the peculiar conditions of aquatic life, although the ancestors of the whale must have lived, like other hair-covered mammals, on land. I cited a number of these transformations--the fish-like form of the body, the hairlessness of the skin, the transformation of the fore-limbs to fins, the disappearance of the hind-limbs and the development of a tail fin, the layer of blubber under the skin, which affords the protection from cold necessary to a warm-blooded animal, the disappearance of the ear-muscles and the auditory passages, the displacement of the external nares to the forehead for the greater security of the breathing-hole during the brief appearance at the surface, and certain remarkable changes in the respiratory and circulatory organs which enable the animal to remain for a long time under water. I might have added many more, for the list of adaptations in the whale to aquatic life is by no means exhausted; they are found in the histological structure and in the minutest combinations in the nervous system. For it is obvious that a tail-fin must be used in quite a different way from a tail, which serves as a fly-brush in hoofed animals, or as an aid to springing in the kangaroo or as a climbing organ; it will require quite different reflex-mechanisms and nerve combinations in the motor centres.

I used this example in order to show how unnecessary it is to assume a special internal evolutionary power for the phylogenesis of species, for this whole order of whales is, so to speak, made up of adaptations; it deviates in many essential respects from the usual mammalian type, and all the deviations are adaptations to aquatic life. But if precisely the most essential features of the organisation thus depend upon adaptation, what is left for a phyletic force to do, since it is these essential features of the structure it would have to determine? There are few people now who believe in a phyletic evolutionary power, which is not made up of the forces known to us--adaptation and heredity--but the conviction that every part of an organism depends upon adaptation has not yet gained a firm footing. Nevertheless, I must continue to regard this conception as the correct one, as I have long done.

I may be permitted one more example. The feather of a bird is a marvellous structure, and no one will deny that as a whole it depends upon adaptation. But what part of it does not depend upon adaptation? The hollow quill, the shaft with its hard, thin, light cortex, and the spongy substance within it, its square section compared with the round section of the quill, the flat barbs, their short, hooked barbules which, in the flight-feathers, hook into one another with just sufficient firmness to resist the pressure of the air at each wing-beat, the lightness and firmness of the whole apparatus, the elasticity of the vane, and so on. And yet all this belongs to an organ which is only passively functional, and therefore can have nothing to do with the Lamarckian principle. Nor can the feather have arisen through some magical effect of temperature, moisture, electricity, or specific nutrition, and thus selection is again our only anchor of safety.

But--it will be objected--the substance of which the feather consists, this peculiar kind of horny substance, did not first arise through selection in the course of the evolution of the birds, for it formed the covering of the scales of their reptilian ancestors. It is quite true that a similar substance covered the scales of the Reptiles, but why should it not have arisen among them through selection? Or in what other way could it have arisen, since scales are also passively useful parts? It is true that if we are only to call adaptation what has been acquired by the species we happen to be considering, there would remain a great deal that could not be referred to selection; but we are postulating an evolution which has stretched back through aeons, and in the course of which innumerable adaptations took place, which had not merely ephemeral persistence in a genus, a family or a class, but which was continued into whole Phyla of animals, with continual fresh adaptations to the special conditions of each species, family, or class, yet with persistence of the fundamental elements. Thus the feather, once acquired, persisted in all birds, and the vertebral column, once gained by adaptation in the lowest forms, has persisted in all the Vertebrates from Amphioxus upwards, although with constant readaptation to the conditions of each particular group. Thus everything we can see in animals is adaptation, whether of to-day, or of yesterday, or of ages long gone by; every kind of cell, whether glandular, muscular, nervous, epidermic, or skeletal, is adapted to absolutely definite and specific functions, and every organ which is composed of these different kinds of cells contains them in the proper proportions, and in the particular arrangement which best serves the function of the organ; it is thus adapted to its function.

All parts of the organism are tuned to one another, that is, they are adapted to one another, and in the same way the organism as a whole is adapted to the conditions of its life, and it is so at every stage of its evolution.

But all adaptations can be referred to selection; the only point that remains doubtful is whether they all must be referred to it.

However that may be, whether the Lamarckian principle is a factor that has been associated with selection in evolution, or whether it is altogether fallacious, the fact remains, that selection is the cause of a great part of the phyletic evolution of organisms on our earth. Those who agree with me in rejecting the Lamarckian principle will regard selection as the only guiding factor in evolution, which creates what is new out of the transmissible variations, by ordering and arranging these, selecting them in relation to their number and size, as the architect does his building-stones so that a particular style must result.[55] But the building-stones themselves, the variations, have their basis in the influences which cause variation in those vital units which are handed on from one generation to another, whether, taken together they form the whole organism, as in Bacteria and other low forms of life, or only a germ-substance, as in unicellular and multicellular organisms.

FOOTNOTES:

[Footnote 33: Vortre Descendenztheorie, Jena, 1904, II. 269. Eng. Transl. London, 1904, II. p. 317.]

[Footnote 34: See Poulton, Essays on Evolution, Oxford, 1908. pp. xix-xxii.]

[Footnote 35: Origin of Species (6th edit), pp. 176 et seq.]

[Footnote 36: Chun, Reise der Valdivia, Leipzig, 1904.]

[Footnote 37: Plate, Selektionsprinzip u. Probleme der Artbildung (3rd edit.), Leipzig, 1908.]

[Footnote 38: Studien zur Descendenz-Theorie II., "Die Enstehung der

Zeichnung bei den Schmetterlings-raupen," Leipzig, 1876.]

[Footnote 39: Origin of Species (6th edit.), p. 232.]

[Footnote 40: Origin of Species, p. 233; see also edit. 1, p. 242.]

[Footnote 41: Ibid. p. 230.]

[Footnote 42: The Effect of External Influences upon Development, Romanes Lecture, Oxford, 1894.]

[Footnote 43: See Poulton, Essays on Evolution, 1908, pp. 316, 317.]

[Footnote 44: The Evolution Theory, London, 1904, I. p. 219.]

[Footnote 45: Report of the British Association (Bristol, 1898), London, 1899, pp. 906-909.]

[Footnote 46: Proc. Ent. Soc., London, May 6, 1903.]

[Footnote 47: Essays on Evolution, 1889-1907, Oxford, 1908, passim, e.g. p. 269.]

[Footnote 48: The expression does not refer to all the enemies of this butterfly; against ichneumon-flies, for instance, their unpleasant smell usually gives no protection.]

[Footnote 49: Professor Poulton has corrected some wrong descriptions which I had unfortunately overlooked in the Plates of my book Vortr 錂 e 黚 er Descendenztheorie, and which refer to Papilio dardanus (merope). These mistakes are of no importance as far as an understanding of the mimicry-theory is concerned, but I hope shortly to be able to correct them in a later edition.]

[Footnote 50: Journ. Linn. Soc. London (Zool.), Vol. xxvi. 1898, pp. 598-602.]

[Footnote 51: In Kosmos, 1879, p. 100.]

[Footnote 52: Habit and Instinct, London. 1896.]

[Footnote 53: This has been discussed in many of my earlier works. See for instance The All-Sufficiency of Natural Selection, a reply to Herbert Spencer, London, 1893.]

[Footnote 54: The Evolution Theory, London, 1904, p. 144.]

[Footnote 55: Variation under Domestication, 1875, II. pp. 426, 427.]

III

HEREDITY AND VARIATION IN MODERN LIGHTS

BY W. BATESON, M.A., F.R.S.

Professor of Biology in the University of Cambridge

Darwin's work has the property of greatness in that it may be admired from more aspects than one. For some the perception of the principle of Natural Selection stands out as his most wonderful achievement to which all the rest is subordinate. Others, among whom I would range myself, look up to him rather as the first who plainly distinguished, collected, and comprehensively studied that new class of evidence from which hereafter a true understanding of the process of Evolution may be developed. We each prefer our own standpoint of admiration; but I think that it will be in their wider aspect that his labours will most command the veneration of posterity.

A treatise written to advance knowledge may be read in two moods. The reader may keep his mind passive, willing merely to receive the impress of the writer's thought; or he may read with his attention strained and alert, asking at every instant how the new knowledge can be used in a further advance, watching continually for fresh footholds by which to climb higher still. Of Shelley it has been said that he was a poet for poets: so Darwin was a naturalist for naturalists. It is when his writings are used in the critical and more exacting spirit with which we test the outfit for our own enterprise that we learn their full value and strength. Whether we glance back and compare his performance with the efforts of his predecessors, or look forward along the course which modern research is disclosing, we shall honour most in him not the rounded merit of finite accomplishment, but the creative power by which he inaugurated a line of discovery endless in variety and extension. Let us attempt thus to see his work in true perspective between the past from which it grew, and the present which is its consequence. Darwin attacked the problem of Evolution by reference to facts of three classes: Variation; Heredity; Natural Selection. His work was not as the laity suppose, a sudden and unheralded revelation, but the first fruit of a long and hitherto barren controversy. The occurrence of variation from type, and the hereditary transmission of such variation had of course been long familiar to practical men, and inferences as to the possible bearing of those phenomena on the nature of specific difference had been from time to time drawn by naturalists

Such passages, of which many (though few so emphatic) can be found in eighteenth century writers, indicate a true perception of the mode of Evolution. The speculations hinted at by Buffon,[57] developed by Erasmus Darwin, and independently proclaimed above all by Lamarck, gave to the doctrine of descent a wide renown. The uniformitarian teaching which Lyell deduced from geological observation had gained acceptance. The facts of geographical distribution[58] had been shown to be obviously inconsistent with the Mosaic legend. Prichard, and Lawrence, following the example of Blumenbach, had successfully demonstrated that the races of Man could be regarded as different forms of one species, contrary to the opinion up till then received. These treatises all begin, it is true, with a profound obeisance to the sons of Noah, but that performed, they continue on strictly modern lines. The question of the mutability of species was thus prominently raised.

Those who rate Lamarck no higher than did Huxley in his contemptuous phrase "buccinator tantum," will scarcely deny that the sound of the trumpet had carried far, or that its note was clear. If then there were few who had already turned to evolution with positive conviction, all scientific men must at least have known that such views had been promulgated; and many must, as Huxley says, have taken up his own position of "critical expectancy."[59]

Why, then, was it, that Darwin succeeded where the rest had failed? The cause of that success was twofold. First, and obviously, in the principle of Natural Selection he had a suggestion which would work. It might not go the whole way, but it was true as far as it went. Evolution could thus in great measure be fairly represented as a consequence of demonstrable processes. Darwin seldom endangers the mechanism he devised by putting on it strains much greater than it can bear. He at least was under no illusion as to the omnipotence of Selection; and he introduces none of the forced pleading which in recent years has threatened to discredit that principle.

For example, in the latest text of the Origin[60] we find him saying:

"But as my conclusions have lately been much misrepresented, and it has been stated that I attribute the modification of species exclusively to natural selection, I may be permitted to remark that in the first edition of this work, and subsequently, I placed in a most conspicuous position--namely, at the close of the Introduction--the following words: 'I am convinced that natural selection has been the main but not the exclusive means of modification.'"

But apart from the invention of this reasonable hypothesis, which may well, as Huxley estimated, "be the guide of biological and psychological speculation for the next three or four generations," Darwin made a more significant and imperishable contribution. Not for a few generations, but through all ages he should be remembered as the first who showed clearly that the problems of Heredity and Variation are soluble by observation, and laid down the course by which we must proceed to their solution.[61] The moment of inspiration did not come with the reading of Malthus, but with the opening of the "first note-book on Transmutation of Species."[62] Evolution is a process of Variation and Heredity. The older writers, though they had some vague idea that it must be so, did not study Variation and Heredity. Darwin did, and so begat not a theory, but a science.

The extent to which this is true, the scientific world is only beginning to realise. So little was the fact appreciated in Darwin's own time that the success of his writings was followed by an almost total cessation of work in that special field. Of the causes which led to these remarkable consequences I have spoken elsewhere. They proceeded from circumstances peculiar to the time; but whatever the causes there is no doubt that this statement of the result is historically exact, and those who make it their business to collect facts elucidating the physiology of Heredity and Variation are well aware that they will find little to reward their quest in the leading scientific Journals of the Darwinian epoch.

In those thirty years the original stock of evidence current and in circulation even underwent a process of attrition. As in the story of the Eastern sage who first wrote the collected learning of the universe for his sons in a thousand volumes and by successive compression and burning reduced them to one and from this by further burning distilled the single ejaculation of the Faith "There is no god but God and Mohammed is the Prophet of God," which was all his maturer wisdom deemed essential:--so in the books of that period do we find the corpus of genetic knowledge dwindle to a few prerogative instances and these at last to the brief formula of an unquestioned creed.

And yet in all else that concerns biological science this period was, in very truth, our Golden Age, when the natural history of the earth was explored as never before; morphology and embryology were exhaustively ransacked; the physiology of plants and animals began to rival chemistry and physics in precision of method and in the rapidity of its advances; and the foundations of pathology were laid.

In contrast with this immense activity elsewhere the neglect which befel the special physiology of Descent, or Genetics as we now call it, is astonishing. This may of course be interpreted as meaning that the favoured studies seemed to promise a quicker return for effort, but it would be more true to say that those who chose these other pursuits did so without making any such comparison; for the idea that the physiology of Heredity and Variation was a coherent science, offering possibilities of extraordinary discovery, was not present to their minds at all. In a word, the existence of such a science was well nigh forgotten. It is true that in ancillary periodicals, as for example those that treat of entomology or horticulture, or in the writings of the already isolated systematists,[63] observations with this special bearing were from time to time related, but the class of fact on which Darwin built his conceptions of Heredity and Variation was not seen in the highways of biology. It formed no part of the official curriculum of biological students, and found no place among the subjects which their teachers were investigating.

During this period nevertheless one distinct advance was made, that with which Weismann's name is prominently connected. In Darwin's genetic scheme the hereditary transmission of parental experience and its consequences played a considerable role. Exactly how great that role was supposed to be, he with his habitual caution refrained from specifying, for the sufficient reason that he did not know. Nevertheless much of the process of Evolution, especially that by which organs have become degenerate and rudimentary, was certainly attributed by Darwin to such inheritance, though since belief in the inheritance of acquired characters fell into dispute, the fact has been a good deal overlooked. The Origin without "use and disuse" would be a materially different book. A certain vacillation is discernible in Darwin's utterances on this question, and the fact gave to the astute Butler an opportunity for his most telling attack. The discussion which best illustrates the genetic views of the period arose in regard to the production of the rudimentary condition of the wings of many beetles in the Madeira group of islands, and by comparing passages from the Origin[64] Butler convicts Darwin of saying first that this condition was in the main the result of Selection, with disuse aiding, and in another place that the main cause of degeneration was disuse, but that Selection had aided. To Darwin however I think the point would have seemed one of dialetics merely. To him the one paramount purpose was to show that somehow an Evolution by means of Variation and Heredity might have brought about the facts observed, and whether they had come to pass in the one way or the other was a matter of subordinate concern.

To us moderns the question at issue has a diminished significance. For over all such debates a change has been brought by Weismann's challenge for evidence that use and disuse have any transmitted effects at all. Hitherto the transmission of many acquired characteristics had seemed to most naturalists so obvious as not to call for demonstration.[65] Weismann's demand for facts in support of the main proposition revealed at once that none having real cogency could be produced. The time-honoured examples were easily shown to be capable of different explanations. A few certainly remain which cannot be so summarily dismissed, but--though it is manifestly impossible here to do justice to such a subject--I think no one will dispute that these residual and doubtful phenomena, whatever be their true nature, are not of a kind to help us much in the interpretation of any of those complex cases of adaptation which on the hypothesis of unguided Natural Selection are especially difficult to understand. Use and disuse were invoked expressly to help us over these hard places; but whatever changes can be induced in offspring by direct treatment of the parents, they are not of a kind to encourage hope of real assistance from that quarter. It is not to be denied that through the collapse of this second line of argument the Selection hypothesis has had to take an increased and perilous burden. Various ways of meeting the difficulty have been proposed, but these mostly resolve themselves into improbable attempts to expand or magnify the powers of Natural Selection.

Weismann's interpellation, though negative in purpose, has had a lasting and beneficial effect, for through his thorough demolition of the old loose and distracting notions of inherited experience, the ground has been cleared for the construction of a true knowledge of heredity based on experimental fact.

In another way he made a contribution of a more positive character, for his elaborate speculations as to the genetic meaning of cytological appearances have led to a minute investigation of the visible phenomena occurring in those cell-divisions by which germ-cells arise. Though the particular views he advocated have very largely proved incompatible with the observed facts of heredity, yet we must acknowledge that it was chiefly through the stimulus of Weismann's ideas that those advances in cytology were made; and though the doctrine of the continuity of germ-plasm cannot be maintained in the form originally propounded, it is in the main true and illuminating.[66] Nevertheless in the present state of knowledge we are still as a rule quite unable to connect cytological appearances with any genetic consequences and save in one respect (obviously of extreme importance--to be spoken of later) the two sets of phenomena might, for all we can see, be entirely distinct.

I cannot avoid attaching importance to this want of connection between the nuclear phenomena and the features of bodily organisation. All attempts to investigate Heredity by cytological means lie under the disadvantage that it is the nuclear changes which can alone be effectively observed. Important as they must surely be, I have never been persuaded that the rest of the cell counts for nothing. What we know of the behaviour and variability of chromosomes seems in my opinion quite incompatible with the belief that they alone govern form, and are the sole agents responsible in heredity.[67]

If, then, progress was to be made in Genetics, work of a different kind was required. To learn the laws of Heredity and Variation there is no other way than that which Darwin himself followed, the direct examination of the phenomena. A beginning could be made by collecting fortuitous observations of this class, which have often thrown a suggestive light, but such evidence can be at best but superficial and some more penetrating instrument of research is required. This can only be provided by actual experiments in breeding.

The truth of these general considerations was becoming gradually clear to many of us when in 1900 Mendel's work was rediscovered. Segregation, a phenomenon of the utmost novelty, was thus revealed. From that moment not only in the problem of the origin of species, but in all the great problems of biology a new era began. So unexpected was the discovery that many naturalists were convinced it was untrue, and at once proclaimed Mendel's conclusions as either altogether mistaken, or if true, of very limited application. Many fantastic notions about the workings of Heredity had been asserted as general principles before: this was probably only another fancy of the same class.

Nevertheless those who had a preliminary acquaintance with the facts of Variation were not wholly unprepared for some such revelation. The essential deduction from the discovery of segregation was that the characters of living things are dependent on the presence of definite elements or factors, which are treated as units in the processes of Heredity. These factors can thus be recombined in various ways. They act sometimes separately, and sometimes they interact in conduction with each other, producing their various effects. All this indicates a definiteness and specific order in heredity, and therefore in variation. This order cannot by the nature of the case be dependent on Natural Selection for its existence, but must be a consequence of the fundamental chemical and physical nature of living things. The study of Variation had from the first shown that an orderliness of this kind was present. The bodies and the properties of livings things are cosmic, not chaotic. No matter how low in the scale we go, never do we find the slightest hint of a diminution in that all-pervading orderliness, nor can we conceive an organism existing for a moment in any other state. Moreover not only does this order prevail in normal forms, but again and again it is to be seen in newly-sprung varieties, which by general consent cannot have been subjected to a prolonged Selection. The discovery of Mendelian elements admirably coincided with and at once gave a rationale of these facts. Genetic Variation is then primarily the consequence of additions to, or omissions from, the stock of elements which the species contains. The further investigation of the species-problem must thus proceed by the analytical method which breeding experiments provide.

In the nine years which have elapsed since Mendel's clue became generally known, progress has been rapid. We now understand the process by which a polymorphic race maintains its polymorphism. When a family consists of dissimilar members, given the numerical proportions in which these members are occurring, we can represent their composition symbolically and state what types can be transmitted by the various members. The difficulty of the "swamping effects of inter-crossing" is practically at an end. Even the famous puzzle of sex-limited inheritance is solved, at all events in its more regular manifestations, and we know now how it is brought about that the normal sisters of a colour-blind man can transmit the colour-blindness while his normal brothers cannot transmit it.

We are still only on the fringe of the inquiry. It can be seen extending and ramifying in many directions. To enumerate these here would be impossible. A whole new range of possibilities is being brought into view by study of the inter-relations between the simple factors. By following up the evidence as to segregation, indications have been obtained which can only be interpreted as meaning that when many factors are being simultaneously redistributed among the germ-cells, certain of them exert what must be described as a repulsion upon other factors. We cannot surmise whither this discovery may lead.

In the new light all the old problems wear a fresh aspect. Upon the question of the nature of Sex, for example, the bearing of Mendelian evidence is close. Elsewhere I have shown that from several sets of parallel experiments the conclusion is almost forced upon us that, in the types investigated, of the two sexes the female is to be regarded as heterozygous in sex, containing one unpaired dominant element, while the male is similarly homozygous in the absence of that element.[68] It is not a little remarkable that on this point-- which is the only one where observations of the nuclear processes of gameto-genesis have yet been brought into relation with the visible characteristics of the organisms themselves--there should be diametrical opposition between the results of breeding experiments and those derived from cytology.

Those who have followed the researches of the American school will be aware that, after it had been found in certain insects that the spermatozoa were of two kinds according as they contained or did not contain the accessory chromosome, E. B. Wilson succeeded in proving that the sperms possessing this accessory body were destined to form females on fertilisation, while sperms without it form males, the eggs being apparently indifferent. Perhaps the most striking of all this series of observations is that lately made by T. H. Morgan,[69] since confirmed by von Baehr, that in a Phylloxeran two kinds of spermatids are formed, respectively with and without an accessory (in this case, double) chromosome. Of these, only those possessing the accessory body become functional spermatozoa, the others degenerating. We have thus an elucidation of the puzzling fact that in these forms fertilisation results in the formation of females only. How the males are formed--for of course males are eventually produced by the parthenogenetic females--we do not know.

If the accessory body is really to be regarded as bearing the factor for femaleness, then in Mendelian terms female is DD and male is DR. The eggs are indifferent and the spermatozoa are each male, or female. But according to the evidence derived from a study of the sex-limited descent of certain features in other animals the conclusion seems equally clear that in them female must be regarded as DR and male as RR. The eggs are thus each either male or female and the spermatozoa are indifferent. How this contradictory evidence is to be reconciled we do not yet know. The breeding work concerns fowls, canaries, and the Currant moth (Abraxas grossulariata). The accessory chromosome has been now observed in most of the great divisions of insects,[70] except, as it happens, Lepidoptera. At first sight it seems difficult to suppose that a feature apparently so fundamental as sex should be differently constituted in different animals, but that seems at present the least improbable inference. I mention these two groups of facts as illustrating the nature and methods of modern genetic work. We must proceed by minute and specific analytical investigation. Wherever we look we find traces of the operation of precise and specific rules.

In the light of present knowledge it is evident that before we can attack the Species-problem with any hope of success there are vast arrears to be made up. He would be a bold man who would now assert that there was no sense in which the term Species might not have a strict and concrete meaning in contradistinction to the term Variety. We have been taught to regard the difference between species and variety as one of degree. I think it unlikely that this conclusion will bear the test of further research. To Darwin the question, What is a variation? presented no difficulties. Any difference between parent and offspring was a variation. Now we have to be more precise. First we must, as de Vries has shown, distinguish real, genetic, variation from fluctuational variations, due to environmental and other accidents, which cannot be transmitted. Having excluded these sources of error the variations observed must be expressed in terms of the factors to which they are due before their significance can be understood. For example, numbers of the variations seen under domestication, and not a few witnessed in nature, are simply the consequence of some ingredient being in an unknown way omitted from the composition of the varying individual. The variation may on the contrary be due to the addition of some new element, but to prove that it is so is by no means an easy matter. Casual observation is useless, for though these latter variations will always be dominants, yet many dominant characteristics may arise from another cause, namely the meeting of complementary factors, and special study of each case in two generations at least is needed before these two phenomena can be distinguished.

When such considerations are fully appreciated it will be realised that medleys of most dissimilar occurrences are all confused together under the term Variation. One of the first objects of genetic analysis is to disentangle this mass of confusion.

To those who have made no study of heredity it sometimes appears that the question of the effect of conditions in causing variation is one which we should immediately investigate, but a little thought will show that before any critical inquiry into such possibilities can be attempted, a knowledge of the working of heredity under conditions as far as possible uniform must be obtained. At the time when Darwin was writing, if a plant brought into cultivation gave off an albino variety, such an event was without hesitation ascribed to the change of life. Now we see that albino gametes, germs, that is to say, which are destitute of the pigment-forming factor, may have been originally produced by individuals standing an indefinite number of generations back in the ancestry of the actual albino, and it is indeed almost certain that the variation to which the appearance of the albino is due cannot have taken place in a generation later than that of the grandparents. It is true that when a new dominant appears we should feel greater confidence that we were witnessing the original variation, but such events are of extreme rarity, and no such case has come under the notice of an experimenter in modern times, as far as I am aware. That they must have appeared is clear enough. Nothing corresponding to the Brown-breasted Game fowl is known wild, yet that colour is a most definite dominant, and at some moment since Gallus bankiva was domesticated, the element on which that special colour depends must have at least once been formed in the germ-cell of a fowl; but we need harder evidence than any which has yet been produced before we can declare that this novelty came through over-feeding, or change of climate, or any other disturbance consequent on domestication. When we reflect on the intricacies of genetic problems as we must now conceive them there come moments when we feel almost thankful that the Mendelian principles were unknown to Darwin. The time called for a bold pronouncement, and he made it, to our lasting profit and delight. With fuller knowledge we pass once more into a period of cautious expectation and reserve.

In every arduous enterprise it is pleasanter to look back at difficulties overcome than forward to those which still seem insurmountable, but in the next stage there is nothing to be stained by disguising the fact that the attributes of living things are not what we used to suppose. If they are more complex in the sense that the properties they display are throughout so regular[71] that the Selection of minute random variations is an unacceptable account of the origin of their diversity, yet by virtue of that very regularity the problem is limited in scope and thus simplified.

To begin with, we must relegate Selection to its proper place. Selection permits the viable to continue and decides that the non-viable shall perish; just as the temperature of our atmosphere decides that no liquid carbon shall be found on the face of the earth: but we do not suppose that the form of the diamond has been gradually achieved by a process of Selection. So again, as the course of descent branches in the successive generations, Selection determines along which branch Evolution shall proceed, but it does not decide what novelties that branch shall bring forth. "La Nature contient le fonds de toutes ces mais le hazard ou l'art les mettent en oeuvre," as Maupertuis most truly said.

Not till knowledge of the genetic properties of organisms has attained to far greater completeness can evolutionary speculations have more than a suggestive value. By genetic experiment, cytology and physiological chemistry aiding, we may hope to acquire such knowledge. In 1872 Nathusius wrote:[72] "Das Gesetz der Vererbung ist noch nicht erkannt; der Apfel ist noch nicht vom Baum der Erkenntniss gefallen, welcher, der Sage nach, Newton auf den rechten Weg zur Ergre dung der Gravitationsgesetze forte." We cannot pretend that the words are not still true, but in Mendelian analysis the seeds of that apple-tree at last are sown.

If we were asked what discovery would do most to forward our inquiry, what one bit of knowledge would more than any other illuminate the problem, I think we may give the answer without hesitation. The greatest advance that we can foresee will be made when it is found possible to connect the geometrical phenomena of development with the chemical. The geometrical symmetry of living things is the key to a knowledge of their regularity, and the forces which cause it. In the symmetry of the dividing cell the basis of that resemblance we call Heredity is contained. To imitate the morphological phenomena of life we have to devise a system which can divide. It must be able to divide, and to segment as--grossly--a vibrating plate or rod does, or as an icicle can do as it becomes ribbed in a continuous stream of water; but with this distinction, that the distribution of chemical differences and properties must simultaneously be decided and disposed in orderly relation to the pattern of the segmentation. Even if a model which would do this could be constructed it might prove to be a useful beginning.

This may be looking too far ahead. If we had to choose some one piece of more proximate knowledge which we would more especially like to acquire, I suppose we should ask for the secret of interracial sterility. Nothing has yet been discovered to remove the grave difficulty, by which Huxley in particular was so much oppressed, that among the many varieties produced under domestication--which we all regard as analogous to the species seen in nature--no clear case of interracial sterility has been demonstrated. The phenomenon is probably the only one to which the domesticated products seem to afford no parallel. No solution of the difficulty can be offered which has positive value, but it is perhaps worth considering the facts in the light of modern ideas. It should be observed that we are not discussing incompatibility of two species to produce offspring (a totally distinct phenomenon), but the sterility of the offspring which many of them do produce.

When two species, both perfectly fertile severally, produce on crossing a sterile progeny, there is a presumption that the sterility is due to the development in the hybrid of some substance which can only be formed by the meeting of two complementary factors. That some such account is correct in essence may be inferred from the well-known observation that if the hybrid is not totally sterile but only partially so, and thus is able to form some good germ-cells which develop into new individuals, the sterility of these daughter-individuals is sensibly reduced or may be entirely absent. The fertility once re-established, the sterility does not return in the later progeny, a fact strongly suggestive of segregation. Now if the sterility of the cross-bred be really the consequence of the meeting of two complementary factors, we see that the phenomenon could only be produced among the divergent offspring of one species by the acquisition of at least two new factors; for if the acquisition of a single factor caused sterility the line would then end. Moreover each factor must be separately acquired by distinct individuals, for if both were present together, the possessors would by hypothesis be sterile. And in order to imitate the case of species each of these factors must be acquired by distinct breeds. The factors need not, and probably would not, produce any other perceptible effects; they might, like the colour-factors present in white flowers, make no difference in the form or other characters. Not till the cross was actually made between the two complementary individuals would either factor come into play, and the effects even then might be unobserved until an attempt was made to breed from the cross-bred.

Next, if the factors responsible for sterility were acquired, they would in all probability be peculiar to certain individuals and would not readily be distributed to the whole breed. Any member of the breed also into which both the factors were introduced would drop out of the pedigree by virtue of its sterility. Hence the evidence that the various domesticated breeds say of dogs or fowls can when mated together produce fertile offspring, is beside the mark. The real question is, Do they ever produce sterile offspring? I think the evidence is clearly that sometimes they do, oftener perhaps than is commonly supposed. These suggestions are quite amenable to experimental tests. The most obvious way to begin is to get a pair of parents which are known to have had any sterile offspring, and to find the proportions in which these steriles were produced. If, as I anticipate, these proportions are found to be definite, the rest is simple.

In passing, certain other considerations may be referred to. First, that there are observations favouring the view that the production of totally sterile cross-breds is seldom a universal property of two species, and that it may be a matter of individuals, which is just what on the view here proposed would be expected. Moreover, as we all know now, though incompatibility may be dependent to some extent on the degree to which the species are dissimilar, no such principle can be demonstrated to determine sterility or fertility in general. For example, though all our Finches can breed together, the hybrids are all sterile. Of Ducks some species can breed together without producing the slightest sterility; others have totally sterile offspring, and so on. The hybrids between several genera of Orchids are perfectly fertile on the female side, and some on the male side also, but the hybrids produced between the Turnip (Brassica napus) and the Swede (Brassica campestris), which, according to our estimates of affinity, should be nearly allied forms, are totally sterile.[73] Lastly, it may be recalled that in sterility we are almost certainly considering a meristic phenomenon. Failure to divide is, we may feel fairly sure, the immediate "cause" of the sterility. Now, though we know very little about the heredity of meristic differences, all that we do know points to the conclusion that the less-divided is dominant to the more-divided, and we are thus justified in supposing that there are factors which can arrest or prevent cell-division. My conjecture therefore is that in the case of sterility of cross-breds we see the effect produced by a complementary pair of such factors. This and many similar problems are now open to our analysis.

The question is sometimes asked, Do the new lights on Variation and Heredity make the process of Evolution easier to understand? On the whole the answer may be given that they do. There is some appearance of loss of simplicity, but the gain is real. As was said above, the time is not ripe for the discussion of the origin of species. With faith in Evolution unshaken--if indeed the word faith can be used in application to that which is certain--we look on the manner and causation of adapted differentiation as still wholly mysterious. As Samuel Butler so truly said: "To me it seems that the 'Origin of Variation,' whatever it is, is the only true 'Origin of Species,'"[74] and of that Origin not one of us knows anything. But given Variation--and it is given: assuming further that the variations are not guided into paths of adaptation--

and both to the Darwinian and to the modern school this hypothesis appears to be sound if unproven--an evolution of species proceeding by definite steps is more, rather than less, easy to imagine than an evolution proceeding by the accumulation of indefinite and insensible steps. Those who have lost themselves in contemplating the miracles of Adaptation (whether real or spurious) have not unnaturally fixed their hopes rather on the indefinite than on the definite changes. The reasons are obvious. By suggesting that the steps through which an adaptative mechanism arose were indefinite and insensible, all further trouble is spared. While it could be said that species arise by an insensible and imperceptible process of variation, there was clearly no use in tiring ourselves by trying to perceive that process. This labour-saving counsel found great favour. All that had to be done to develop evolution-theory was to discover the good in everything, a task which, in the complete absence of any control or test whereby to check the truth of the discovery, is not very onerous. The doctrine "que tout est au mieux" was therefore preached with fresh vigour, and examples of that illuminating principle were discovered with a facility that Pangloss himself might have envied, till at last even the spectators wearied of such dazzling performances.

But in all seriousness, why should indefinite and unlimited variation have been regarded as a more probable account of the origin of Adaptation? Only, I think, because the obstacle was shifted one plane back, and so looked rather less prominent. The abundance of Adaptation, we all grant, is an immense, almost an unsurpassable difficulty in all non-Lamarckian views of Evolution; but if the steps by which that adaptation arose were fortuitous, to imagine them insensible is assuredly no help. In one most important respect indeed, as has often been observed, it is a multiplication of troubles. For the smaller the steps, the less could Natural Selection act upon them. Definite variations--and of the occurrence of definite variations in abundance we have now the most convincing proof--have at least the obvious merit that they can make and often do make a real difference in the chances of life.

There is another aspect of the Adaptation problem to which I can allude very briefly. May not our present ideas of the universality and precision of Adaptation be greatly exaggerated? The fit of organism to its environment is not after all so very close--a proposition unwelcome perhaps, but one which could be illustrated by very copious evidence. Natural Selection is stern, but she has her tolerant moods.

We have now most certain and irrefragable proof that much definiteness exists in living things apart from Selection, and also much that may very well have been preserved and so in a sense constituted by Selection. Here the matter is likely to rest. There is a passage in the sixth edition of the Origin which has I think been overlooked. On page 70 Darwin says, "The tuft of hair on the breast of the wild turkey-cock cannot be of any use, and it is doubtful whether it can be ornamental in the eyes of the female bird." This tuft of hair is a most definite and unusual structure, and I am afraid that the remark that it "cannot be of any use" may have been made inadvertently; but it may have been intended, for in the first edition the usual qualification was given and must therefore have been deliberately excised. Anyhow I should like to think that Darwin did throw over that tuft of hair, and that he felt relief when he had done so. Whether however we have his great authority for such a course or not, I feel quite sure that we shall be rightly interpreting the facts of nature if we cease to expect to find purposefulness wherever we meet with definite structures or patterns. Such things are, as often as not, I suspect rather of the nature of tool-marks, mere incidents of manufacture, benefiting their possessor not more than the wire-marks in a sheet of paper, or the ribbing on the bottom of an oriental plate renders those objects more attractive in our eyes.

If Variation may be in any way definite, the question once more arises, may it not be definite in direction? The belief that it is has had many supporters, from Lamarck onwards, who held that it was guided by need, and others who, like Neli, while laying no emphasis on need, yet were convinced that there was guidance of some kind. The latter view under the name of "Orthogenesis," devised I believe by Eimer, at the present day commends itself to some naturalists. The objection to such a suggestion is of course that no fragment of real evidence can be produced in its support. On the other hand, with the experimental proof that variation consists largely in the unpacking and repacking of an original complexity, it is not so certain as we might like to think that the order of these events is not predetermined.

For instance the original "pack" may have been made in such a way that at the nth division of the germ-cells of a Sweet Pea a colour-factor might be dropped, and that at the n+nth division the hooded variety be given off, and so on. I see no ground whatever for holding such a view, but in fairness the possibility should not be forgotten, and in the light of modern research it scarcely looks so absurdly improbable as before.

No one can survey the work of recent years without perceiving that evolutionary orthodoxy developed too fast, and that a great deal has got to come down; but this satisfaction at least remains, that in the experimental methods which Mendel inaugurated, we have means of reaching certainty in regard to the physiology of Heredity and Variation upon which a more lasting structure may be built.

FOOTNOTES:

[Footnote 56: Vous Physique, contenant deux Dissertations, l'une sur l'origine des Hommes et des Animaux; Et l'autre sur l'origine des Noirs, La Haye, 1746, pp. 124 and 129. For an introduction to the writings of Maupertuis I am indebted to an article by Professor Lovejoy in Popular Sci. Monthly, 1902.]

[Footnote 57: For the fullest account of the views of these pioneers of Evolution, see the works of Samuel Butler, especially Evolution, Old and New (2nd edit.) 1882. Butler's claims on behalf of Buffon have met with some acceptance; but after reading what Butler has said, and a considerable part of Buffon's own works, the word "hinted" seems to me a sufficiently correct description of the part he played.

[Footnote 58: See especially W. Lawrence, Lectures on Physiology, London, 1823, pp. 213 f.]

[Footnote 59: See the chapter contributed to the Life and Letters of Charles Darwin, II. p. 195. I do not clearly understand the sense in which Darwin wrote (Autobiography, ibid. I. p. 87): "It has sometimes been said that the success of the Origin proved 'that the subject was in the air,' or 'that men's minds were prepared for it.' I do not think that this is strictly true, for I occasionally sounded not a few naturalists, and never happened to come across a single one who seemed to doubt about the permanence of species." This experience may perhaps have been an accident due to Darwin's isolation. The literature of the period abounds with indications of "critical expectancy." A most interesting expression of that feeling is given in the charming account of the "Early Days of Darwinism" by Alfred Newton, Macmillan's Magazine, LVII. 1888, p. 241. He tells how in 1858 when spending a dreary summer in Iceland, he and his friend, the ornithologist John Wolley, in default of active occupation, spent their days in discussion. "Both of us taking a keen interest in Natural History, it was but reasonable that a question, which in those days was always coming up wherever two or more naturalists were gathered together, should be continually recurring. That question was, 'What is a species?' and connected therewith was the other question, 'How did a species begin?'... Now we were of course fairly well acquainted with what had been published on these subjects." He then enumerates some of these publications, mentioning among others T. Vernon Wollaston's Variation of Species--a work which has in my opinion never been adequately appreciated. He proceeds: "Of course we never arrived at anything like a solution of these problems, general or special, but we felt very strongly that a solution ought to be found, and that quickly, if the study of Botany and Zoology was to make any great advance." He then describes how on his return home he received the famous number of the Linnean Journal on a certain evening. "I sat up late that night to read it; and never shall I forget the impression it made upon me. Herein was contained a perfectly simple solution of all the difficulties which had been troubling me for months past.... I went to bed satisfied that a solution had been found."]

[Footnote 60: Origin, 6th edit. (1882), p. 421.]

[Footnote 61: Whatever be our estimate of the importance of Natural Selection, in this we all agree. Samuel Butler, the most brilliant, and by far the most interesting of Darwin's opponents--whose works are at length emerging from oblivion--in his Preface (1882) to the 2nd edition of Evolution, Old and New, repeats his earlier expression of homage to one whom he had come to regard as an enemy: "To the end of time, if the question be asked, 'Who taught people to believe in Evolution?' the answer must be that it was Mr. Darwin. This is true, and it is hard to see what palm of higher praise can be awarded to any philosopher."]

[Footnote 62: Life and Letters, I. pp. 276 and 83.]

[Footnote 63: This isolation of the systematists is the one most melancholy sequela of Darwinism. It seems an irony that we should read in the peroration to the Origin that when the Darwinian view is accepted "Systematists will be able to pursue their labours as at present; but they will not be incessantly haunted by the shadowy doubt whether this or that form be a true species. This, I feel sure, and I speak after experience, will be no slight relief. The endless disputes whether or not some fifty species of British brambles are good species will cease." Origin, 6th edit. (1882), p. 425. True they have ceased to attract the attention of those who lead opinion, but anyone who will turn to the literature of systematics will find that they have not ceased in any other sense. Should there not be something disquieting in the fact that among the workers who come most into contact with specific differences, are to be found the only men who have failed to be persuaded of the unreality of those differences?]

[Footnote 64: 6th edit. pp. 109 and 401. See Butler, Essays on Life, Art, and Science, p. 265, reprinted 1908, and Evolution, Old and New, chap. XXII. (2nd edit.), 1882.]

[Footnote 65: W. Lawrence was one of the few who consistently maintained the contrary opinion. Prichard, who previously had expressed himself in the same sense, does not, I believe, repeat these views in his later writings, and there are signs that he came to believe in the transmission of acquired habits. See Lawrence, Lect. Physiol. 1823, pp. 436-437, 447. Prichard, Edin. Inaug. Disp. 1808 [not seen by me], quoted ibid. and Nat. Hist. Man, 1843, pp. 34 f.]

[Footnote 66: It is interesting to see how nearly Butler was led by natural penetration, and from absolutely opposite conclusions, back to this underlying truth: "So that each ovum when impregnate should be considered not as descended from its ancestors, but as being a continuation of the personality of every ovum in the chain of its ancestry, which every ovum it actually is quite as truly as the octogenarian is the same identity with the ovum from which he has been developed. This process cannot stop short of the primordial cell, which again will probably turn out to be but a brief resting-place. We therefore prove each one of us to be actually the primordial cell which never died nor dies, but has differentiated itself into the life of the world, all living beings whatever, being one with it and members one of another," Life and Habit, 1878, p. 86.]

[Footnote 67: This view is no doubt contrary to the received opinion. I am however interested to see it lately maintained by Driesch (Science and Philosophy of the Organism, London, 1907, p. 233), and from the recent observations of Godlewski it has received distinct experimental support.]

[Footnote 68: In other words, the ova are each either female, or male (i.e. non-female), but the sperms are all non-female.]

[Footnote 69: Morgan, Proc. Soc. Exp. Biol. Med. V. 1908, and von Baehr, Zool. Anz. XXXII. p. 507, 1908.]

[Footnote 70: As Wilson has proved, the unpaired body is not a universal feature even in those orders in which it has been observed. Nearly allied types may differ. In some it is altogether unpaired. In others it is paired with a body of much smaller size, and by selection of various types all gradations can be demonstrated ranging to the condition in which the members of the pair are indistinguishable from each other.]

[Footnote 71: I have in view, for example, the marvellous and specific phenomena of regeneration, and those discovered by the students of "Entwicklungsmechanik." The circumstances of its occurrence here preclude any suggestion that this regularity has been brought about by the workings of Selection. The attempts thus to represent the phenomena have resulted in mere parodies of scientific reasoning.]

[Footnote 72: Vortre Viehzucht und Rassenerkenntniss, p. 120, Berlin, 1872.]

[Footnote 73: See Sutton, A. W., Journ. Linn. Soc. XXXVIII. p. 341, 1908.]

[Footnote 74: Life and Habit, London, p. 263, 1878]

IV

"THE DESCENT OF MAN"

BY G. SCHWALBE

Professor of Anatomy in the University of Strassburg

The problem of the origin of the human race, of the descent of man, is ranked by Huxley in his epoch-making book Man's Place in Nature, as the deepest with which biology has to concern itself, "the question of questions,"--the problem which underlies all others. In the same brilliant and lucid exposition, which appeared in 1863, soon after the publication of Darwin's Origin of Species, Huxley stated his own views in regard to this great problem. He tells us how the idea of a natural descent of man gradually grew up in his mind. It was especially the assertions of Owen in regard to the total difference between the human and the simian brain that called forth strong dissent from the great anatomist Huxley, and he easily succeeded in showing that Owen's supposed differences had no real existence; he even established, on the basis of his own anatomical investigations, the proposition that the anatomical differences between the Marmoset and the Chimpanzee are much greater than those between the Chimpanzee and Man.

But why do we thus introduce the study of Darwin's Descent of Man, which is to occupy us here, by insisting on the fact that Huxley had taken the field in defence of the descent of man in 1863, while Darwin's book on the subject did not appear till 1871? It is in order that we may clearly understand how it happened that from this time onwards Darwin and Huxley followed the same great aim in the most intimate association.

Huxley and Darwin working at the same Problema maximum! Huxley fiery, impetuous, eager for battle, contemptuous of the resistance of a dull world, or energetically triumphing over it. Darwin calm, weighing every problem slowly, letting it mature thoroughly,--not a fighter, yet having the greater and more lasting influence by virtue of his immense mass of critically sifted proofs. Darwin's friend, Huxley, was the first to do him justice, to understand his nature, and to find in it the reason why the detailed and carefully considered book on the descent of man made its appearance so late. Huxley, always generous, never thought of claiming priority for himself. In enthusiastic language he tells how Darwin's immortal work, The Origin of Species, first shed light for him on the problem of the descent of man; the recognition of a vera causa in the transformation of species illuminated his thoughts as with a flash. He was now content to leave what perplexed him, what he could not yet solve, as he says himself, "in the mighty hands of Darwin." Happy in the bustle of strife against old and deep-rooted prejudices, against intolerance and superstition, he wielded his sharp weapons on Darwin's behalf; wearing

Darwin's armour he joyously overthrew adversary after adversary. Darwin spoke of Huxley as his "general agent."[75] Huxley says of himself "I am Darwin's bulldog."[76]

Thus Huxley openly acknowledged that it was Darwin's Origin of Species that first set the problem of the descent of man in its true light, that made the question of the origin of the human race a pressing one. That this was the logical consequence of his book Darwin himself had long felt. He had been reproached with intentionally shirking the application of his theory to Man. Let us hear what he says on this point in his autobiography: "As soon as I had become, in the year 1837 or 1838, convinced that species were mutable productions, I could not avoid the belief that man must come under the same law. Accordingly I collected notes on the subject for my own satisfaction, and not for a long time with any intention of publishing. Although in the 'Origin of Species' the derivation of any particular species is never discussed, yet I thought it best, in order that no honourable man should accuse me of concealing my views,[77] to add that by the work 'light would be thrown on the origin of man and his history.' It would have been useless and injurious to the success of the book to have paraded, without giving any evidence, my conviction with respect to his origin."[78]

In a letter written in January, 1860, to the Rev. L. Blomefield, Darwin expresses himself in similar terms. "With respect to man, I am very far from wishing to obtrude my belief; but I thought it dishonest to quite conceal my opinion."[79]

The brief allusion in the Origin of Species is so far from prominent and so incidental that it was excusable to assume that Darwin had not touched upon the descent of man in this work. It was solely the desire to have his mass of evidence sufficiently complete, solely Darwin's great characteristic of never publishing till he had carefully weighed all aspects of his subject for years, solely, in short, his most fastidious scientific conscience that restrained him from challenging the world in 1859 with a book in which the theory of the descent of man was fully set forth. Three years, frequently interrupted by ill-health, were needed for the actual writing of the book:[80] the first edition, which appeared in 1871, was followed in 1874 by a much improved second edition, the preparation of which he very reluctantly undertook.[81]

This, briefly, is the history of the work, which, with the Origin of Species, marks an epoch in the history of biological sciences--the work with which the cautious, peace-loving investigator ventured forth from his contemplative life into the arena of strife and unrest, and laid himself open to all the annoyances that deep-rooted belief and prejudice, and the prevailing tendency of scientific thought at the time could devise.

Darwin did not take this step lightly. Of great interest in this connection is a letter written to Wallace on Dec. 22, 1857,[82] in which he says, "You ask me whether I shall discuss 'man.' I think I shall avoid the whole subject, as so surrounded with prejudices; though I fully admit that it is the highest and most interesting problem for the naturalist." But his conscientiousness compelled him to state briefly his opinion on the subject in the Origin of Species in 1859. Nevertheless he did not escape reproaches for having been so reticent. This is unmistakably apparent from a letter to Fritz M黮ler dated Feb. 22 [1869?], in which he says: "I am thinking of writing a little essay on the Origin of Mankind, as I have been taunted with concealing my opinions."[83]

It might be thought that Darwin behaved thus hesitatingly, and was so slow in deciding on the full publication of his collected material in regard to the descent of man, because he had religious difficulties to overcome.

But this was not the case, as we can see from his admirable confession of faith, the publication of which we owe to his son Francis.[84] Whoever wishes really to understand the lofty character of this great man should read these immortal lines in which he unfolds to us in simple and straightforward words the development of his conception of the universe. He describes how, though he was still quite orthodox during his voyage round the world on board the Beagle, he came gradually to see, shortly afterwards (1836-1839) that the Old Testament was no more to be trusted than the Sacred Books of the Hindoos; the miracles by which Christianity is supported, the discrepancies between the accounts in the different Gospels, gradually led him to disbelieve in Christianity as a divine revelation. "Thus," he writes,[85] "disbelief crept over me at a very slow rate, but was at last complete. The rate was so slow that I felt no distress." But Darwin was too modest to presume to go beyond the limits laid down by science. He wanted nothing more than to be able to go, freely and unhampered by belief in authority or in the Bible, as far as human knowledge could lead him. We learn this from the concluding words of his chapter on religion "The mystery of the beginning of all things is insoluble by us; and I for one must be content to remain an Agnostic."[86]

Darwin was always very unwilling to give publicity to his views in regard to religion. In a letter to Asa Gray on May 22, 1860,[87] he declares that it is always painful to him to have to enter into discussion of religious problems. He had, he said, no intention of writing atheistically.

Finally, let us cite one characteristic sentence from a letter from Darwin to C. Ridley[88] (Nov. 28, 1878). A clergyman, Dr. Pusey, had asserted that Darwin had written the Origin of Species with some relation to theology. Darwin writes emphatically, "Many years ago when I was collecting facts for the 'Origin,' my belief in what is called a personal God was as firm as that of Dr. Pusey himself, and as to the eternity of matter I never troubled myself about such insoluble questions." The expression "many years ago" refers to the time of his voyage round the world, as has already been pointed out. Darwin means by this utterance that the views which had gradually developed in his mind in regard to the origin of species were quite compatible with the faith of the Church.

If we consider all these utterances of Darwin in regard to religion and to his outlook on life (Weltanschauung), we shall see at least so much, that religious reflection could in no way have influenced him in regard to the writing and publishing of his book on The Descent of Man. Darwin had early won for himself freedom of thought, and to this freedom he remained true to the end of his life, uninfluenced by the customs and opinions of the world around him.

Darwin was thus inwardly fortified and armed against the host of calumnies, accusations, and attacks called forth by the publication of the Origin of Species, and to an even greater extent by the appearance of the Descent of Man. But in his defence he could rely on the aid of a band of distinguished auxiliaries of the rarest ability. His faithful confederate, Huxley, was joined by the botanist Hooker, and, after longer resistance, by the famous geologist Lyell, whose "conversion" afforded Darwin peculiar satisfaction. All three took the field with enthusiasm in defence of the natural descent of man. From Wallace, on the other hand, though he shared with him the idea of natural selection, Darwin got no support in this matter. Wallace expressed himself in a strange manner. He admitted everything in regard to the morphological descent of man, but maintained, in a mystic way, that something else, something of a spiritual nature must have been added to what man inherited from his animal ancestors. Darwin, whose esteem for Wallace was extraordinarily high, could not understand how he could give utterance to such a mystical view in regard to man; the idea seemed to him so "incredibly strange" that he thought some one else must have added these sentences to Wallace's paper.

Even now there are thinkers who, like Wallace, shrink from applying to man the ultimate consequences of the theory of descent. The idea that man is derived from ape-like forms is to them unpleasant and humiliating.

So far I have been depicting the development of Darwin's work on the descent of man. In what follows I shall endeavour to give a condensed survey of the contents of the book.

It must at once be said that the contents of Darwin's work fall into two parts, dealing with entirely different subjects. The Descent of Man includes a very detailed investigation in regard to secondary sexual characters in the animal series, and on this investigation Darwin founded a new theory, that of sexual selection. With astonishing patience he gathered together an immense mass of material, and showed, in regard to Arthropods and Vertebrates, the wide distribution of secondary characters, which develop almost exclusively in the male, and which enable him, on the one hand, to get the better of his rivals in the struggle for the female by the greater perfection of his weapons, and, on the other hand, to offer greater allurements to the female through the higher development of decorative characters, of song, or of scent-producing glands. The best equipped males will thus crowd out the less well-equipped in the matter of reproduction, and thus the relevant characters will be increased and perfected through sexual selection. It is, of course, a necessary assumption that these secondary sexual characters may be transmitted to the female, although perhaps in rudimentary form.

As we have said, this story of sexual selection takes up a great deal of space in Darwin's book, and it need only be considered here in so far as Darwin applied it to the descent of man. To this latter problem the whole of Part I is devoted, while Part III contains a discussion of sexual selection in relation to man, and a general summary. Part II treats of sexual selection in general, and may be disregarded in our present study. Moreover, many interesting details must necessarily be passed over in what follows, for want of space.

The first part of the Descent of Man begins with an enumeration of the proofs of the animal descent of man taken from the structure of the human body. Darwin chiefly emphasises the fact that the human body consists of the same organs and of the same tissues as those of the other mammals; he shows also that man is subject to the same diseases and tormented by the same parasites as the apes. He further dwells on the general agreement exhibited by young embryonic forms, and he illustrates this by two figures placed one above the other, one representing a human embryo, after Ecker, the other a dog embryo, after Bischoff.[89]

Darwin finds further proofs of the animal origin of man in the reduced structures, in themselves extremely variable, which are either absolutely useless to their possessors, or of so little use that they could never have developed under existing conditions. Of such vestiges he enumerates: the defective development of the panniculus carnosus (muscle of the skin) so widely distributed among mammals, the ear-muscles, the occasional persistence of the animal ear-point in man, the rudimentary nictitating membrane (plica semilunaris) in the human eye, the slight development of the organ of smell, the general hairiness of the human body, the frequently defective development or entire absence of the third molar (the wisdom tooth), the vermiform appendix, the occasional reappearance of a bony canal (foramen supracondyloideum) at the lower end of the humerus, the rudimentary tail of man (the so-called taillessness), and so on. Of these rudimentary structures the occasional occurrence of the animal ear-point in man is most fully discussed. Darwin's attention was called to this interesting structure by the sculptor Woolner. He figures such a case observed in man, and also the head of an alleged orang-foetus, the photograph of which he received from Nitsche.

Darwin's interpretation of Woolner's case as having arisen through a folding over of the free edge of a pointed ear has been fully borne out by my investigations on the external ear.[90] In particular, it was established by these investigations that the human foetus, about the middle of its embryonic life, possesses a pointed ear somewhat similar to that of the monkey genus Macacus. One of Darwin's statements in regard to the head of the orang-foetus must be corrected. A large ear with a point is shown in the photograph,[91] but it can easily be demonstrated--and Deniker has already pointed this out--that the figure is not that of an orang foetus at all, for that form has much smaller ears with no point; nor can it be a gibbon-foetus, as Deniker supposes, for the gibbon ear is also without a point. I myself regard it as that of a Macacus-embryo. But this mistake, which is due to Nitsche, in no way affects the fact recognised by Darwin, that ear-forms showing the point characteristic of the animal ear occur in man with extraordinary frequency.

Finally, there is a discussion of those rudimentary structures which occur only in one sex, such as the rudimentary mammary glands in the male, the vesicula prostatica, which corresponds to the uterus of the female, and others. All these facts tell in favour of the common descent of man and all other vertebrates. The conclusion of this section is characteristic: "It is only our natural prejudice, and that arrogance which made our forefathers declare that they were descended from demi-gods, which leads us to demur to this conclusion. But the time will before long come, when it will be thought wonderful that naturalists, who were well acquainted with the comparative structure and development of man, and other mammals, should have believed that each was the work of a separate act of creation."[92]

In the second chapter there is a more detailed discussion, again based upon an extraordinary wealth of facts, of the problem as to the manner in which, and the causes through which, man evolved from a lower form. Precisely the same causes are here suggested for the origin of man, as for the origin of species in general. Variability, which is a necessary assumption in regard to all transformations, occurs in man to a high degree. Moreover, the rapid multiplication of the human race creates conditions which necessitate an energetic struggle for existence, and thus afford scope for the intervention of natural selection. Of the exercise of artificial selection in the human race, there is nothing to be said, unless we cite such cases as the grenadiers of Frederick William I, or the population of ancient Sparta. In the passages already referred to and in those which follow, the transmission of acquired characters, upon which Darwin does not dwell, is taken for granted. In man, direct effects of changed conditions can be demonstrated (for instance in regard to bodily size), and there are also proofs of the influence exerted on his physical constitution by increased use or disuse. Reference is here made to the fact, established by Forbes, that the Quechua Indians of the high plateaus of Peru show a striking development of lungs and thorax, as a result of living constantly at high altitudes.

Such special forms of variation as arrests of development (microcephalism) and reversion to lower forms are next discussed. Darwin himself felt[93] that these subjects are so nearly related to the cases mentioned in the first chapter, that many of them might as well have been dealt with there. It seems to me that it would have been better so, for the citation of additional instances of reversion at this place rather disturbs the logical sequence of his ideas as to the conditions which have brought about the evolution of man from lower forms. The instances of reversion here discussed are microcephalism, which Darwin wrongly interpreted as atavistic, supernumerary mammae, supernumerary digits, bicornuate uterus, the development of abnormal muscles, and so on. Brief mention is also made of correlative variations observed in man.

Darwin next discusses the question as to the manner in which man attained to the erect position from the state of a climbing quadruped. Here again he puts the influence of Natural Selection in the first rank. The immediate progenitors of man had to maintain a struggle for existence in which success was to the more intelligent, and to those with social instincts. The hand of these climbing ancestors, which had little skill and served mainly for locomotion, could only undergo further development when some early member of the Primate series came to live more on the ground and less among trees.

A bipedal existence thus became possible, and with it the liberation of the hand from locomotion, and the one-sided development of the human foot. The upright position brought about correlated variations in the bodily structure; with the free use of the hand it became possible to manufacture weapons and to use them; and this again resulted in a degeneration of the powerful canine teeth and the jaws, which were then no longer necessary for defence. Above all, however, the intelligence immediately increased, and with it skull and brain. The nakedness of man, and the absence of a tail (rudimentariness of the tail vertebrae) are next discussed. Darwin is inclined to attribute the nakedness of man, not to the action of natural selection on ancestors who originally inhabited a tropical land, but to sexual selection, which, for aesthetic reasons, brought about the loss of the hairy covering in man, or primarily in woman. An interesting discussion of the loss of the tail, which, however, man shares with the anthropoid apes, some other monkeys and lemurs, forms the conclusion of the almost superabundant material which Darwin worked up in the second chapter. His object was to show that some of the most distinctive human characters are in all probability directly or indirectly due to natural selection. With characteristic modesty he adds:[94] "Hence, if I have erred in giving to natural selection great power, which I am very far from admitting, or in having exaggerated its power, which is in itself probable, I have at least, as I hope, done good service in aiding to overthrow the dogma of separate creations." At the end of the chapter he touches upon the objection as to man's helpless and defenceless condition. Against this he urges his intelligence and social instincts.

The two following chapters contain a detailed discussion of the objections drawn from the supposed great differences between the mental powers of men and animals. Darwin at once admits that the differences are enormous, but not that any fundamental difference between the two can be found. Very characteristic of him is the following passage: "In what manner the mental powers were first developed in the lowest organisms, is as hopeless an enquiry as how life itself first originated. These are problems for the distant future, if they are ever to be solved by man."[95]

After some brief observations on instinct and intelligence, Darwin brings forward evidence to show that the greater number of the emotional states, such as pleasure and pain, happiness and misery, love and hate are common to man and the higher animals. He goes on to give various examples showing that wonder and curiosity, imitation, attention, memory and imagination (dreams of animals), can also be observed in the higher mammals, especially in apes. In regard even to reason there are no sharply defined limits. A certain faculty of deliberation is characteristic of some animals, and the more thoroughly we know an animal the more intelligence we are inclined to credit it with. Examples are brought forward of the intelligent and deliberate actions of apes, dogs and elephants. But although no sharply defined differences exist between man and animals, there is, nevertheless, a series of other mental powers which are characteristics usually regarded as absolutely peculiar to man. Some of these characteristics are examined in detail, and it is shown that the arguments drawn from them are not conclusive. Man alone is said to be capable of progressive improvement; but against this must be placed as something analogous in animals, the fact that they learn cunning and caution through long continued persecution. Even the use of tools is not in itself peculiar to man (monkeys use sticks, stones and twigs), but man alone fashions and uses implements designed for a special purpose. In this connection the remarks taken from Lubbock in regard to the origin and gradual development of the earliest flint implements will be read with interest; these are similar to the observations on modern eoliths, and their bearing on the development of the stone industry. It is interesting to learn from a letter to Hooker,[96] that Darwin himself at first doubted whether the stone implements discovered by Boucher de Perthes were really of the nature of tools. With the relentless candour as to himself which characterised him, he writes four years later in a letter to Lyell in regard to this view of Boucher de Perthes' discoveries: "I know something about his errors, and looked at his book many years ago, and am ashamed to think that I concluded the whole was rubbish! Yet he has done for man something like what Agassiz did for glaciers."[97]

To return to Darwin's further comparisons between the higher mental powers of man and animals; He takes much of the force from the argument that man alone is capable of abstraction and self-consciousness by his own observations on dogs. One of the main differences between man and animals, speech, receives detailed treatment. He points out that various animals (birds, monkeys, dogs) have a large number of different sounds for different emotions, that, further, man produces in common with animals a whole series of inarticulate cries combined with gestures, and that dogs learn to understand whole sentences of human speech. In regard to human language, Darwin expresses a view contrary to that held by Max Müler:[98] "I cannot doubt that language owes its origin to the imitation and modification of various natural sounds, the voices of other animals, and man's own instinctive cries, aided by signs and gestures." The development of actual language presupposes a higher degree of intelligence than is found in any kind of ape. Darwin remarks on this point:[99] "The fact of the higher apes not using their vocal organs for speech no doubt depends on their intelligence not having been sufficiently advanced."

The sense of beauty, too, has been alleged to be peculiar to man. In refutation of this assertion Darwin points to the decorative colours of birds, which are used for display. And to the last objection, that man alone has religion, that he alone has a belief in God, it is answered "that numerous races have existed, and still exist, who have no idea of one or more gods, and who have no words in their languages to express such an idea."[100]

The result of the investigations recorded in this chapter is to show that, great as the difference in mental powers between man and the higher animals may be, it is undoubtedly only a difference "of degree and not of kind."[101]

In the fourth chapter Darwin deals with the moral sense or conscience, which is the most important of all differences between man and animals. It is a result of social instincts, which lead to sympathy for other members of the same society, to non-egoistic actions for the good of others. Darwin shows that social tendencies are found among many animals, and that among these love and kin-sympathy exist, and he gives examples of animals (especially dogs) which may exhibit characters that we should call moral in man (e.g. disinterested self-sacrifice for the sake of others). The early ape-like progenitors of the human race were undoubtedly social. With the increase of intelligence the moral sense develops farther; with the acquisition of speech public opinion arises, and finally, moral sense becomes habit. The rest of Darwin's detailed discussions on moral philosophy may be passed over.

The fifth chapter may be very briefly summarised. In it Darwin shows that the intellectual and moral faculties are perfected through natural selection. He inquires how it can come about that a tribe at a low level of evolution attains to a higher, although the best and bravest among them often pay for their fidelity and courage with their lives without leaving any descendants. In this case it is the sentiment of glory, praise and blame, the admiration of others, which bring about the increase of the better members of the tribe. Property, fixed dwellings, and the association of families into a community are also indispensable requirements for civilisation. In the longer second section of the fifth chapter Darwin acts mainly as recorder. On the basis of numerous investigations, especially those of Greg, Wallace, and Galton, he inquires how far the influence of natural selection can be demonstrated in regard to civilised nations. In the final section, which deals with the proofs that all civilised nations were once barbarians, Darwin again uses the results gained by other investigators, such as Lubbock and Tylor. There are two sets of facts which prove the proposition in question. In the first place, we find traces of a former lower state in the customs and beliefs of all civilised nations, and in the second place, there are proofs to show that savage races are independently able to raise themselves a few steps in the scale of civilisation, and that they have thus raised themselves.

In the sixth chapter of the work, Morphology comes into the foreground once more. Darwin first goes back, however, to the argument based on the great difference between the mental powers of the highest animals and those of man. That this is only quantitative, not qualitative, he has already shown. Very instructive in this connection is the reference to the enormous difference in mental powers in another class. No one would draw from the fact that the cochineal insect (Coccus) and the ant exhibit enormous differences in their mental powers, the conclusion that the ant should therefore be regarded as something quite distinct, and withdrawn from the class of insects altogether.

Darwin next attempts to establish the specific genealogical tree of man, and carefully weighs the differences and resemblances between the different families of the Primates. The erect position of man is an adaptive character, just as are the various characters referable to aquatic life in the seals, which, notwithstanding these, are ranked as a mere family of the carnivores. The following utterance is very characteristic of Darwin:[102] "If man had not been his own classifier, he would never have thought of founding a separate order for his own reception." In numerous characters not mentioned in systematic works, in the features of the face, in the form of the nose, in the structure of the external ear, man resembles the apes. The arrangement of the hair in man has also much in common with the apes; as also the occurrence of hair on the forehead of the human embryo, the beard, the convergence of the hair of the upper and under arm towards the elbow, which occurs not only in the anthropoid apes, but also in some American monkeys. Darwin here adopts Wallace's explanation of the origin of the ascending direction of the hair in the forearm of the orang,--that it has arisen through the habit of holding the hands over the head in rain. But this explanation cannot be maintained when we consider that this disposition of the hair is widely distributed among the most different mammals, being found in the dog, in the sloth, and in many of the lower monkeys.

After further careful analysis of the anatomical characters Darwin reaches the conclusion that the New World monkeys (Platyrrhine) may be excluded from the genealogical tree altogether, but that man is an offshoot from the Old World monkeys (Catarrhine) whose progenitors existed as far back as the Miocene period. Among these Old World monkeys the forms to which man shows the greatest resemblance are the anthropoid apes, which, like him, possess neither tail nor ischial callosities. The platyrrhine and catarrhine monkeys have their primitive ancestor among extinct forms of the Lemuridae. Darwin also touches on the question of the original home of the human race and supposes that it may have been in Africa, because it is there that man's nearest relatives, the gorilla and the chimpanzee, are found. But he regards speculation on this point as useless. It is remarkable that, in this connection, Darwin regards the loss of the hair-covering in man as having some relation to a warm climate, while elsewhere he is inclined to make sexual selection responsible for it. Darwin recognises the great gap between man and his nearest relatives, but similar gaps exist at other parts of the mammalian genealogical tree: the allied forms have become extinct. After the extermination of the lower races of mankind, on the one hand, and of the anthropoid apes on the other, which will undoubtedly take place, the gulf will be greater than ever, since the baboons will then bound it on the one side, and the white races on the other. Little weight need be attached to the lack of fossil remains to fill up this gap, since the discovery of these depends upon chance. The last part of the chapter is devoted to a discussion of the earlier stages in the genealogy of man. Here Darwin accepts in the main the genealogical tree, which had meantime been published by Haeckel, who traces the pedigree back through Monotrems, Reptiles, Amphibians, and Fishes, to Amphioxus.

Then follows an attempt to reconstruct, from the atavistic characters, a picture of our primitive ancestor who was undoubtedly an arboreal animal. The occurrence of rudiments of parts in one sex which only come to full development in the other is next discussed. This state of things Darwin regards as derived from an original hermaphroditism. In regard to the mammary glands of the male he does not accept the theory that they are vestigial, but considers them rather as not fully developed.

The last chapter of Part I deals with the question whether the different races of man are to be regarded as different species, or as sub-species of a race of monophyletic origin. The striking differences between the races are first emphasised, and the question of the fertility or infertility of hybrids is discussed. That fertility is the more usual is shown by the excessive fertility of the hybrid population of Brazil. This, and the great variability of the distinguishing characters of the different races, as well as the fact that all grades of transition stages are found between these, while considerable general agreement exists, tell in favour of the unity of the races and lead to the conclusion that they all had a common primitive ancestor.

Darwin therefore classifies all the different races as sub-species of one and the same species. Then follows an interesting inquiry into the reasons for the extinction of human races. He recognises as the ultimate reason the injurious effects of a change of the conditions of life, which may bring about an increase in infantile mortality, and a diminished fertility. It is precisely the reproductive system, among animals also, which is most susceptible to changes in the environment.

The final section of this chapter deals with the formation of the races of mankind. Darwin discusses the question how far the direct effect of different conditions of life, or the inherited effects of increased use or disuse may have brought about the characteristic differences between the different races. Even in regard to the origin of the colour of the skin he rejects the transmitted effects of an original difference of climate as an explanation. In so doing he is following his tendency to exclude Lamarckian explanations as far as possible. But here he makes gratuitous difficulties from which, since natural selection fails, there is no escape except by bringing in the principle of sexual selection, to which, he regarded it as possible, skin-colouring, arrangement of hair, and form of features might be traced. But with his characteristic conscientiousness he guards himself thus: "I do not intend to assert that sexual selection will account for all the differences between the races."[103]

I may be permitted a remark as to Darwin's attitude towards Lamarck. While, at an earlier stage, when he was engaged in the preliminary labours for his immortal work, The Origin of Species, Darwin expresses himself very forcibly against the views of Lamarck, speaking of Lamarckian "nonsense,"[104] and of Lamarck's "absurd, though clever work"[105] and expressly declaring, "I attribute very little to the direct action of climate, etc."[106] yet in later life he became more and more convinced of the influence of external conditions. In 1876, that is, two years after the appearance of the second edition of The Descent of Man, he writes with his usual candid honesty: "In my opinion the greatest error which I have committed, has been not allowing sufficient weight to the direct action of the environment, i.e. food, climate, etc. independently of a natural selection."[107] It is certain from this change of opinion that, if he had been able to make up his mind to issue a third edition of The Descent of Man, he would have ascribed a much greater influence to the effect of external conditions in explaining the different characters of the races of man than he did in the second edition. He would also undoubtedly have attributed less influence to sexual selection as a factor in the origin of the different bodily characteristics, if indeed he would not have excluded it altogether.

In Part III of the Descent two additional chapters are devoted to the discussion of sexual selection in relation to man. These may be very briefly referred to. Darwin here seeks to show that sexual selection has been operative on man and his primitive progenitor. Space fails me to follow out his interesting arguments. I can only mention that he is inclined to trace back hairlessness, the development of the beard in man, and the characteristic colour of the different human races to sexual selection. Since bareness of the skin could be no advantage, but rather a disadvantage, this character cannot have been brought about by natural selection. Darwin also rejected a direct influence of climate as a cause of the origin of the skin-colour. I have already expressed the opinion, based on the development of his views as shown in his letters, that in a third edition Darwin would probably have laid more stress on the influence of external environment. He himself feels that there are gaps in his proofs here, and says in self-criticism: "The views here advanced, on the part which sexual selection has played in the history of man, want scientific precision."[108] I need here only point out that it is impossible to explain the graduated stages of skin-colour by sexual selection, since it would have produced races sharply defined by their colour and not united to other races by transition stages, and this, it is well known, is not the case. Moreover, the fact established by me,[109] that in all races the ventral side of the trunk is paler than the dorsal side, and the inner surface of the extremities paler than the outer side, cannot be explained by sexual selection in the Darwinian sense.

With this I conclude my brief survey of the rich contents of Darwin's book. I may be permitted to conclude by quoting the magnificent final words of The Descent of Man: "We must, however, acknowledge, as it seems to me, that man, with all his noble qualities, with sympathy which feels for the most debased, with benevolence which extends not only to other men but to the humblest living creature, with his god-like intellect which has penetrated into the movements and constitution of the solar system--with all these exalted powers--Man still bears in his bodily frame the indelible stamp of his lowly origin."[110]

What has been the fate of Darwin's doctrines since his great achievement? How have they been received and followed up by the scientific and lay world? And what do the successors of the mighty hero and genius think now in regard to the origin of the human race?

At the present time we are incomparably more favourably placed than Darwin was for answering this question of all questions. We have at our command an incomparably greater wealth of material than he had at his disposal. And we are more fortunate than he in this respect, that we now know transition-forms which help to fill up the gap, still great, between the lowest human races and the highest apes. Let us consider for a little the more essential additions to our knowledge since the publication of The Descent of Man.

Since that time our knowledge of animal embryos has increased enormously. While Darwin was obliged to content himself with comparing a human embryo with that of a dog, there are now available the youngest embryos of monkeys of all possible groups (Orang, Gibbon, Semnopithecus, Macacus), thanks to Selenka's most successful tour in the East Indies in search of such material. We can now compare corresponding stages of the lower monkeys and of the Anthropoid apes with human embryos, and convince ourselves of their great resemblance to one another, thus strengthening enormously the armour prepared by Darwin in defence of his view on man's nearest relatives. It may be said that Selenka's material fills up the blanks in Darwin's array of proofs in the most satisfactory manner.

The deepening of our knowledge of comparative anatomy also gives us much surer foundations than those on which Darwin was obliged to build.

Just of late there have been many workers in the domain of the anatomy of apes and lemurs, and their investigations extend to the most different organs. Our knowledge of fossil apes and lemurs has also become much wider and more exact since Darwin's time: the fossil lemurs have been especially worked up by Cope, Forsyth Major, Ameghino, and others. Darwin knew very little about fossil monkeys. He mentions two or three anthropoid apes as occurring in the Miocene of Europe,[111] but only names Dryopithecus, the largest form from the Miocene of France. It was erroneously supposed that this form was related to Hylobates. We now know not only a form that actually stands near to the gibbon (Pliopithecus), and remains of other anthropoids (Pliohylobates and the fossil chimpanzee, Palaeopithecus), but also several lower catarrhine monkeys, of which Mesopithecus, a form nearly related to the modern Sacred Monkeys (a species of Semnopithecus) and found in strata of the Miocene period in Greece, is the most important. Quite recently, too, Ameghino's investigations have made us acquainted with fossil monkeys from South America (Anthropops, Homunculus), which, according to their discoverer, are to be regarded as in the line of human descent.

What Darwin missed most of all--intermediate forms between apes and man--has been recently furnished. E. Dubois, as is well known, discovered in 1893, near Trinil in Java, in the alluvial deposits of the river Bengawan, an important form represented by a skull-cap, some molars, and a femur. His opinion--much disputed as it has been--that in this form, which he named Pithecanthropus, he has found a long-desired transition-form is shared by the present writer. And although the geological age of these fossils, which, according to Dubois, belong to the uppermost Tertiary series, the Pliocene has recently been fixed at a later date (the older Diluvium), the morphological value of these interesting remains, that is, the intermediate position of Pithecanthropus, still holds good. Volz says with justice,[112] that even if Pithecanthropus is not the missing link, it is undoubtedly a missing link.

As on the one hand there has been found in Pithecanthropus a form which, though intermediate between apes and man, is nevertheless more closely allied to the apes, so on the other hand, much progress has been made since Darwin's day in the discovery and description of the oldest human remains. Since the famous roof of a skull and the bones of the extremities belonging to it were found in 1856 in the Neandertal near Dusseldorf, the most varied judgments have been expressed in regard to the significance of the remains and of the skull in particular. In Darwin's Descent of Man there is only a passing allusion to them[113] in connection with the discussion of the skull-capacity, although the investigations of Schaaffhausen, King, and Huxley were then known. I believe I have shown, in a series of papers, that the skull in question belongs to a form different from any of the races of man now living, and, with King and Cope, I regard it as at least a different species from living man, and have therefore designated it Homo primigenius. The form unquestionably belongs to the older Diluvium, and in the later Diluvium human forms already appear, which agree in all essential points with existing human races.

As far back as 1886 the value of the Neandertal skull was greatly enhanced by Fraipont's discovery of two skulls and skeletons from Spy in Belgium. These are excellently described by their discoverer,[114] and are regarded as belonging to the same group of forms as the Neandertal remains. In 1899 and the following years came the discovery by Gorjanovic-Kramberger of different skeletal parts of at least ten individuals in a cave near Krapina in Croatia.[115] It is in particular the form of the lower jaw which is different from that of all recent races of man, and which clearly indicates the lowly position of Homo primigenius, while, on the other hand, the long-known skull from Gibraltar, which I[116] have referred to Homo primigenius, and which has lately been examined in detail by Sollas,[117] has made us acquainted with the surprising shape of the eye-orbit, of the nose, and of the whole upper part of the face. Isolated lower jaws found at La Naulette in Belgium, and at Malarnaud in France, increase our material which is now as abundant as could be desired. The most recent discovery of all is that of a skull dug up in August of this year [1908] by Klaatsch and Hauser in the lower grotto of the Le Moustier in Southern France, but this skull has not yet been fully described. Thus Homo primigenius must also be regarded as occupying a position in the gap existing between the highest apes and the lowest human races, Pithecanthropus, standing in the lower part of it, and Homo primigenius in the higher, near man. In order to prevent misunderstanding, I should like here to emphasise that in arranging this structural series--anthropoid apes, Pithecanthropus, Homo primigenius, Homo sapiens--I have no intention of establishing it as a direct genealogical series. I shall have something to say in regard to the genetic relations of these forms, one to another, when discussing the different theories of descent current at the present day.[118]

In quite a different domain from that of morphological relationship, namely in the physiological study of the blood, results have recently been gained which are of the highest importance to the doctrine of descent. Uhlenhuth, Nuttall, and others have established the fact that the blood-serum of a rabbit which has previously had human blood injected into it, forms a precipitate with human blood. This biological reaction was tried with a great variety of mammalian species, and it was found that those far removed from man gave no precipitate under these conditions. But as in other cases among mammals all nearly related forms yield an almost equally marked precipitate, so the serum of a rabbit treated with human blood and then added to the blood of an anthropoid ape gives almost as marked a precipitate as in human blood; the reaction to the blood of the lower Eastern monkeys is weaker, that to the Western monkeys weaker still; indeed in this last case there is only a slight clouding after a considerable time and no actual precipitate. The blood of the Lemuridae (Nuttall) gives no reaction or an extremely weak one, that of the other mammals none whatever. We have in this not only a proof of the literal blood relationship between man and apes, but the degree of relationship with the different main groups of apes can be determined beyond possibility of mistake.

Finally, it must be briefly mentioned that in regard to remains of human handicraft also, the material at our disposal has greatly increased of late years, that, as a result of this, the opinions of archaeologists have undergone many changes, and that, in particular, their views in regard to the age of the human race have been greatly influenced. There is a tendency at the present time to refer the origin of man back to Tertiary times. It is true that no remains of Tertiary man have been found, but flints have been discovered which, according to the opinion of most investigators, bear traces either of use, or of very primitive workmanship. Since Rutot's time, following Mortillet's example, investigators have called these "eoliths," and they have been traced back by Verworn to the Miocene of the Auvergne, and by Rutot even to the upper Oligocene. Although these eoliths are even nowadays the subject of many different views, the preoccupation with them has kept the problem of the age of the human race continually before us.

Geology, too, has made great progress since the days of Darwin and Lyell, and has endeavoured with satisfactory results to arrange the human remains of the Diluvial period in chronological order (Penck). I do not intend to enter upon the question of the primitive home of the human race; since the space at my disposal will not allow of my touching even very briefly upon all the departments of science which are concerned in the problem of the descent of man. How Darwin would have rejoiced over each of the discoveries here briefly outlined! What use he would have made of the new and precious material, which would have prevented the discouragement from which he suffered when preparing the second edition of The Descent of Man! But it was not granted to him to see this progress towards filling up the gaps in his edifice of which he was so painfully conscious.

He did, however, have the satisfaction of seeing his ideas steadily gaining ground, notwithstanding much hostility and deep-rooted prejudice. Even in the years between the appearance of The Origin of Species and of the first edition of the Descent, the idea of a natural descent of man, which was only briefly indicated in the work of 1859, had been eagerly welcomed in some quarters. It has been already pointed out how brilliantly Huxley contributed to the defence and diffusion of Darwin's doctrines, and how in Man's Place in Nature he has given us a classic work as a foundation for the doctrine of the descent of man. As Huxley was Darwin's champion in England, so in Germany Carl Vogt, in particular, made himself master of the Darwinian ideas. But above all it was Haeckel who, in energy, eagerness for battle, and knowledge may be placed side by side with Huxley, who took over the leadership in the controversy over the new conception of the universe. As far back as 1866, in his Generelle Morphologie, he had inquired minutely into the question of the descent of man, and not content with urging merely the general theory of descent from lower animal forms, he drew up for the first time genealogical trees showing the close structural relationships of the different animal groups; the last of these illustrated the relationships of Mammals, and among them of all groups of the Primates, including man. It was Haeckel's genealogical trees that formed the basis of the special discussion of the relationships of man, in the sixth chapter of Darwin's Descent of Man.

In the last section of this essay I shall return to Haeckel's conception of the special descent of man, the main features of which he still upholds, and rightly so. Haeckel has contributed more than any one else to the spread of the Darwinian doctrine.

I can only allow myself a few words as to the spread of the theory of the natural descent of man in other countries. The Parisian anthropological school, founded and guided by the genius of Broca, took up the idea of the descent of man, and made many notable contributions to it (Broca, Manouvrier, Mahoudeau, Deniker and others). In England itself Darwin's work did not die. Huxley took care of that, for he, with his lofty and unprejudiced mind, dominated and inspired English biology until his death on June 29, 1895. He had the satisfaction shortly before his death of learning of Dubois' discovery, which he illustrated by a humourous sketch.[119] But there are still many followers in Darwin's footsteps in England. Keane has worked at the special genealogical tree of the Primates; Keith has inquired which of the anthropoid apes has the greatest number of characters in common with man; Morris concerns himself with the evolution of man in general, especially with his acquisition of the erect position. The recent discoveries of Pithecanthropus and Homo primigenius are being vigorously discussed; but the present writer is not in a position to form an opinion of the extent to which the idea of descent has penetrated throughout England generally.

In Italy independent work in the domain of the descent of man is being produced, especially by Morselli; with him are associated, in the investigation of related problems, Sergi and Giuffrida-Ruggeri. From the ranks of American investigators we may single out in particular the eminent geologist Cope, who championed with much decision the idea of the specific difference of Homo neandertalensis (primigenius) and maintained a more direct descent of man from the fossil Lemuridae. In South America too, in Argentina, new life is stirring in this department of science. Ameghino in Buenos Ayres has awakened the fossil primates of the Pampas formation to new life; he even believes that in his Tetraprothomo, represented by a femur, he has discovered a direct ancestor of man. Lehmann-Nitsche is working at the other side of the gulf between apes and man, and he describes a remarkable first cervical vertebra (atlas) from Monte Hermoso as belonging to a form which may bear the same relation to Homo sapiens in South America as Homo primigenius does in the Old World. After a minute investigation he establishes a human species Homo neogaeus, while Ameghino ascribes this atlas vertebra to his Tetraprothomo.

Thus throughout the whole scientific world there is arising a new life, an eager endeavour to get nearer to Huxley's problema maximum, to penetrate more deeply into the origin of the human race. There are to-day very few experts in anatomy and zoology who deny the animal descent of man in general. Religious considerations, old prejudices, the reluctance to accept man, who so far surpasses mentally all other creatures, as descended from "soulless" animals, prevent a few investigators from giving full adherence to the doctrine. But there are very few of these who still postulate a special act of creation for man. Although the majority of experts in anatomy and zoology accept unconditionally the descent of man from lower forms, there is much diversity of opinion among them in regard to the special line of descent.

In trying to establish any special hypothesis of descent, whether by the graphic method of drawing up genealogical trees or otherwise, let us always bear in mind Darwin's words[120] and use them as a critical guiding line: "As we have no record of the lines of descent, the pedigree can be discovered only by observing the degrees of resemblance between the beings which are to be classed." Darwin carries this further by stating "that resemblances in several unimportant structures, in useless and rudimentary organs, or not now functionally active, or in an embryological condition, are by far the most serviceable for classification."[121] It has also to be remembered that numerous separate points of agreement are of much greater importance than the amount of similarity or dissimilarity in a few points.

The hypotheses as to descent current at the present day may be divided into two main groups. The first group seeks for the roots of the human race not among any of the families of the apes--the anatomically nearest forms--nor among their very similar but less specialised ancestral forms, the fossil representatives of which we can know only in part, but, setting the monkeys on one side, it seeks for them lower down among the fossil Eocene Pseudo-lemuridae or Lemuridae (Cope), or even among the primitive pentadactylous Eocene forms, which may either have led directly to the evolution of man (Adloff), or have given rise to an ancestral form common to apes and men (Klaatsch,[122] Giuffrida-Ruggeri). The common ancestral form, from which man and apes are thus supposed to have arisen independently, may explain the numerous resemblances which actually exist between them. That is to say, all the characters upon which the great structural resemblance between apes and man depends must have been present in their common ancestor. Let us take an example of such a common character. The bony external ear-passage is in general as highly developed in the lower Eastern monkeys and the anthropoid apes as in man. This character must, therefore, have already been present in the common primitive form. In that case it is not easy to understand why the Western monkeys have not also inherited the character, instead of possessing only a tympanic ring. But it becomes more intelligible if we assume that forms with a primitive tympanic ring were the original type, and that from these were evolved, on the one hand, the existing New World monkeys with persistent tympanic ring, and on the other an ancestral form common to the lower Old World monkeys, the anthropoid apes and man. For man shares with these the character in question, and it is also one of the "unimportant" characters required by Darwin. Thus we have two divergent lines arising from the ancestral form, the Western monkeys (Platyrrhine) on the one hand, and an ancestral form common to the lower Eastern monkeys, the anthropoid apes, and man, on the other. But considerations similar to those which showed it to be impossible that man should have developed from an ancestor common to him and the monkeys, yet outside of and parallel with these, may be urged also against the likelihood of a parallel evolution of the lower Eastern monkeys, the anthropoid apes, and man. The anthropoid apes have in common with man many characters which are not present in the lower Old World monkeys. These characters must therefore have been present in the ancestral form common to the three groups. But here, again, it is difficult to understand why the lower Eastern monkeys should not also have inherited these characters. As this is not the case, there remains no alternative but to assume divergent evolution from an indifferent form. The lower Eastern monkeys are carrying on the evolution in one direction--I might almost say towards a blind alley--while anthropoids and men have struck out a progressive path, at first in common, which explains the many points of resemblance between them, without regarding man as derived directly from the anthropoids. Their many striking points of agreement indicate a common descent, and cannot be explained as phenomena of convergence.

I believe I have shown in the above sketch that a theory which derives man directly from lower forms without regarding apes as transition-types leads ad absurdum. The close structural relationship between man and monkeys can only be understood if both are brought into the same line of evolution. To trace man's line of descent directly back to the old Eocene mammals, alongside of, but with no relation to these very similar forms, is to abandon the method of exact comparison, which, as Darwin rightly recognised, alone justifies us in drawing up genealogical trees on the basis of resemblances and differences. The farther down we go the more does the ground slip from beneath our feet. Even the Lemuridae show very numerous divergent conditions, much more so the Eocene mammals (Creodonta, Condylarthra), the chief resemblance of which to man consists in the possession of pentadactylous hands and feet! Thus the farther course of the line of descent disappears in the darkness of the ancestry of the mammals. With just as much reason we might pass by the Vertebrates altogether, and go back to the lower Invertebrates, but in that case it would be much easier to say that man has arisen independently, and has evolved, without relation to any animals, from the lowest primitive form to his present isolated and dominant position. But this would be to deny all value to classification, which must after all be the ultimate basis of a genealogical tree. We can, as Darwin rightly observed, only infer the line of descent from the degree of resemblance between single forms. If we regard man as directly derived from primitive forms very far back, we have no way of explaining the many points of agreement between him and the monkeys in general, and the anthropoid apes in particular. These must remain an inexplicable marvel.

I have thus, I trust, shown that the first class of special theories of descent, which assumes that man has developed, parallel with the monkeys, but without relation to them, from very low primitive forms cannot be upheld, because it fails to take into account the close structural affinity of man and monkeys. I cannot but regard this hypothesis as lamentably retrograde, for it makes impossible any application of the facts that have been discovered in the course of the anatomical and embryological study of man and monkeys, and indeed prejudges investigations of that class as pointless. The whole method is perverted; an unjustifiable theory of descent is first formulated with the aid of the imagination, and then we are asked to declare that all structural relations between man and monkeys, and between the different groups of the latter, are valueless,--the fact being that they are the only true basis on which a genealogical tree can be constructed.

So much for this most modern method of classification, which has probably found adherents because it would deliver us from the relationship to apes which many people so much dislike. In contrast to it we have the second class of special hypotheses of descent, which keeps strictly to the nearest structural relationship. This is the only basis that justifies the drawing up of a special hypothesis of descent. If this fundamental proposition be recognised, it will be admitted that the doctrine of special descent upheld by Haeckel, and set forth in Darwin's Descent of Man, is still valid to-day. In the genealogical tree, man's place is quite close to the anthropoid apes; these again have as their nearest relatives the lower Old World monkeys, and their progenitors must be sought among the less differentiated Platyrrhine monkeys, whose most important characters have been handed on to the present day New World monkeys. How the different genera are to be arranged within the general scheme indicated depends in the main on the classificatory value attributed to individual characters. This is particularly true in regard to Pithecanthropus, which I consider as the root of a branch which has sprung from the anthropoid ape root and has led up to man; the latter I have designated the family of the Hominidae.

For the rest, there are, as we have said, various possible ways of constructing the narrower genealogy within the limits of this branch including men and apes, and these methods will probably continue to change with the accumulation of new facts. Haeckel himself has modified his genealogical tree of the Primates in certain details since the publication of his Generelle Morphologie in 1866, but its general basis remains the same.[123] All the special genealogical trees drawn up on the lines laid down by Haeckel and Darwin--and that of Dubois may be specially mentioned--are based, in general, on the close relationship of monkeys and men, although they may vary in detail. Various hypotheses have been formulated on these lines, with special reference to the evolution of man. Pithecanthropus is regarded by some authorities as the direct ancestor of man, by others as a side-track failure in the attempt at the evolution of man. The problem of the monophyletic or polyphyletic origin of the human race has also been much discussed. Sergi[124] inclines towards the assumption of a polyphyletic origin of the three main races of man, the African primitive form of which has given rise also to the gorilla and chimpanzee, the Asiatic to the Orang, the Gibbon, and Pithecanthropus. Kollmann regards existing human races as derived from small primitive races (pigmies), and considers that Homo primigenius must have arisen in a secondary and degenerative manner.

But this is not the place, nor have I the space to criticise the various special theories of descent. One, however, must receive particular notice. According to Ameghino, the South American monkeys (Pitheculites) from the oldest

Tertiary of the Pampas are the forms from which have arisen the existing American monkeys on the one hand, and on the other, the extinct South American Homunculidae, which are also small forms. From these last, anthropoid apes and man have, he believes, been evolved. Among the progenitors of man, Ameghino reckons the form discovered by him (Tetraprothomo), from which a South American primitive man, Homo pampaeus, might be directly evolved, while on the other hand all the lower Old World monkeys may have arisen from older fossil South American forms (Clenialitidae), the distribution of which may be explained by the bridge formerly existing between South America and Africa, as may be the derivation of all existing human races from Homo pampaeus.[125] The fossil forms discovered by Ameghino deserve the most minute investigation, as does also the fossil man from South America of which Lehmann-Nitsche[126] has made a thorough study.

It is obvious that, notwithstanding the necessity for fitting man's line of descent into the genealogical tree of the Primates, especially the apes, opinions in regard to it differ greatly in detail. This could not be otherwise, since the different Primate forms, especially the fossile forms, are still far from being exhaustively known. But one thing remains certain,--the idea of the close relationship between man and monkeys set forth in Darwin's Descent of Man. Only those who deny the many points of agreement, the sole basis of classification, and thus of a natural genealogical tree, can look upon the position of Darwin and Haeckel as antiquated, or as standing on an insufficient foundation. For such a genealogical tree is nothing more than a summarised representation of what is known in regard to the degree of resemblance between the different forms.

Darwin's work in regard to the descent of man has not been surpassed; the more we immerse ourselves in the study of the structural relationships between apes and man, the more is our path illumined by the clear light radiating from him, and through his calm and deliberate investigation, based on a mass of material in the accumulation of which he has never had an equal. Darwin's fame will be bound up for all time with the unprejudiced investigation of the question of all questions, the descent of the human race.

FOOTNOTES:

[Footnote 75: Life and Letters of Thomas Henry Huxley, Vol. I. p. 171, London, 1900.]

[Footnote 76: Ibid., p. 363.]

[Footnote 77: No italics in original.]

[Footnote 78: Life and Letters of Charles Darwin, Vol. I. p. 93.]

[Footnote 79: Ibid. Vol. II. p. 263.]

[Footnote 80: Ibid. Vol. I. p. 94.]

[Footnote 81: Life and Letters, Vol. III. p. 175.]

[Footnote 82: Ibid. Vol. II. p. 109.]

[Footnote 83: Ibid. Vol. III. p. 112.]

[Footnote 84: Ibid. Vol. I. pp. 304-317.]

[Footnote 85: Life and Letters, Vol. I. p. 309.]

[Footnote 86: Loc. cit. p. 313.]

[Footnote 87: Ibid. Vol. II. p. 310.]

[Footnote 88: Ibid. Vol. III. p. 236. ["C. Ridley," Mr. Francis Darwin points out to me, should be H. N. Ridley. A.C.S.]]

[Footnote 89: Descent of Man (Popular Edit., 1901), fig. 1, p. 14.]

[Footnote 90: G. Schwalbe, "Das Darwin'sche Spitzohr beim menschlichen Embryo," Anatom. Anzeiger, 1889, pp. 176-189, and other papers.]

[Footnote 91: Descent of Man, fig. 3, p. 24.]

[Footnote 92: Descent of man, p. 6.]

[Footnote 93: Ibid. p. 54.]

[Footnote 94: Descent of Man, p. 92.]

[Footnote 95: Ibid. p. 100.]

[Footnote 96: Life and letters, Vol. II. p. 161, June 22, 1859.]

[Footnote 97: Ibid. Vol. III. p. 15, March 17, 1863.]

[Footnote 98: Descent of Man, p. 132.]

[Footnote 99: Ibid. pp. 136, 137.]

[Footnote 100: Ibid. p. 143.]

[Footnote 101: Ibid. p. 193.]

[Footnote 102: Descent of Man, p. 231.]

[Footnote 103: Descent of Man, p. 308.]

[Footnote 104: Life and Letters, Vol. II. p. 23.]

[Footnote 105: Loc. cit. p. 39.]

[Footnote 106: Loc. cit. (1856), p. 82.]

[Footnote 107: Ibid. Vol. III p. 159.]

[Footnote 108: Descent of Man, p. 924.]

[Footnote 109: "Die Hautfarbe des Menschen," Mitteilungen der Anthropologischen Gessellschaft in Wien, Vol. XXXIV. pp. 331-352.]

[Footnote 110: Ibid. p. 947.]

[Footnote 111: Descent of Man, p. 240.]

[Footnote 112: "Das geologische Alter der Pithecanthropus-Schichten bei Trinil, Ost-Java." Neues Jahrb. f. Mineralogie. Festband, 1907.]

[Footnote 113: Descent of Man, p. 82.]

[Footnote 114: "La race humaine de Neanderthal ou de Canstatt en Belgique." Arch. de Biologie, VII. 1887.]

[Footnote 115: Gorjanovic-Kramberger. Der diluviale Mensch van Krapina in Kroatien, 1906.]

[Footnote 116: Studien zur Vorgeschichte des Menschen, 1906, pp. 154 ff.]

[Footnote 117: "On the cranial and facial characters of the Neandertal Race." Trans. R. Soc. London, vol. 199, 1908, p. 281.]

[Footnote 118: Since this essay was written Schoetensack has discovered near Heidelberg and briefly described an exceedingly interesting lower jaw from rocks between the Pliocene and Diluvial beds. This exhibits interesting differences from the forms of lower jaw of Homo primigenius. (Schoetensack, Der Unterkiefer des Homo heidelbergensis, Leipzig, 1908.) G. S.]

[Footnote 119: Life and Letters of Thomas Henry Huxley, Vol. II. p. 394.]

[Footnote 120: Descent of Man, p. 229.]

[Footnote 121: Loc. cit.]

[Footnote 122: Klaatsch in his last publications speaks in the main only of an ancestral form common to men and anthropoid apes.]

[Footnote 123: Haeckels latest genealogical tree is to be found in his most recent work, Unsere Ahnenreihe. Jena, 1908.]

[Footnote 124: Sergi, G. Europa, 1908.]

[Footnote 125: See Ameghino's latest paper, "Notas preliminaries sobre el Tetraprothomo argentinus," etc. Anales del Museo nacional de Buenos Aires, XVI. pp. 107-242, 1907.]

[Footnote 126: "Nouvelles recherches sur la formation pamponne et l'homme fossile de la Republique Argentine." Rivista del Museo de la Plata, T. XIV. pp. 193-488.]

V

CHARLES DARWIN AS AN ANTHROPOLOGIST

BY ERNST HAECKEL

Professor of Zoology in the University of Jena

The great advance that anthropology has made in the second half of the nineteenth century is due, in the first place, to Darwin's discovery of the origin of man. No other problem in the whole field of research is so momentous as that of "Man's place in nature," which was justly described by Huxley (1863) as the most fundamental of all questions. Yet the scientific solution of this problem was impossible until the theory of descent had been established.

It is now a hundred years since the great French biologist Jean Lamarck published his Philosophie Zoologique. By a remarkable coincidence the year in which that work was issued, 1809, was the year of the birth of his most distinguished successor, Charles Darwin. Lamarck had already recognised that the descent of man from a series of other Vertebrates--that is, from a series of Ape-like Primates--was essentially involved in the general theory of transformation which he had erected on a broad inductive basis; and he had sufficient penetration to detect the agencies that had been at work in the evolution of the erect bimanous man from the arboreal and quadrumanous ape. He had, however, few empirical arguments to advance in support of his hypothesis, and it could not be established until the further development of the biological sciences--the founding of comparative embryology by Baer (1828) and of the cell-theory by Schleiden and Schwann (1838), the advance of physiology under Johannes Muller (1833), and the enormous progress of palaeontology and comparative anatomy between 1820 and 1860--provided this necessary foundation. Darwin was the first to coordinate the ample results of these lines of research. With no less comprehensiveness than discrimination he consolidated them as a basis of a modified theory of descent, and associated with them his own theory of natural selection, which we take to be distinctive of "Darwinism" in the stricter sense. The illuminating truth of these cumulative arguments was so great in every branch of biology that, in spite of the most vehement opposition, the battle was won within a single decade, and Darwin secured the general admiration and recognition that had been denied to his forerunner, Lamarck, up to the hour of his death (1829).

Before, however, we consider the momentous influence that Darwinism has had in anthropology, we shall find it useful to glance at its history in the course of the last half century, and notice the various theories that have contributed to its advance. The first attempt to give extensive expression to the reform of biology by Darwin's work will be found in my Generelle Morphologie (1866)[127] which was followed by a more popular treatment of the subject in my Nat 黯 liche Sch 鮨 fungsgeschichte (1868),[128] a compilation from the earlier work. In the first volume of the Generelle Morphologie I endeavoured to show the great importance of evolution in settling the fundamental questions of biological philosophy, especially in regard to comparative anatomy. In the second volume I dealt broadly with the principle of evolution, distinguishing ontogeny and phylogeny as its two coordinate main branches, and associating the two in the Biogenetic Law. The Law may be formulated thus: "Ontogeny (embryology or the development of the individual) is a concise and compressed recapitulation of phylogeny (the palaeontological or genealogical series) conditioned by laws of heredity and adaptation." The "Systematic introduction to general evolution," with which the second volume of the Generelle Morphologie opens, was the first attempt to draw up a natural system of organisms (in harmony with the principles of Lamarck and Darwin) in the form of a hypothetical pedigree, and was provisionally set forth in eight genealogical tables.

In the nineteenth chapter of the Generelle Morphologie--a part of which has been republished, without any alteration, after a lapse of forty years--I made a critical study of Lamarck's theory of descent and of Darwin's theory of selection, and endeavoured to bring the complex phenomena of heredity and adaptation under definite laws for the first time. Heredity I divided into conservative and progressive: adaptation into indirect (or potential) and direct (or actual). I then found it possible to give some explanation of the correlation of the two physiological functions in the struggle for life (selection), and to indicate the important laws of divergence (or differentiation) and complexity (or division of labor), which are the direct and inevitable outcome of selection. Finally, I marked off dysteleology as the science of the aimless (vestigial, abortive, atrophied, and useless) organs and parts of the body. In all this I worked from a strictly monistic standpoint, and sought to explain all biological phenomena on the mechanical and naturalistic lines that had long been recognised in the study of inorganic nature. Then (1866), as now, being convinced of the unity of nature, the fundamental identity of the agencies at work in the inorganic and the organic worlds, I discarded vitalism, teleology, and all hypotheses of a mystic character.

It was clear from the first that it was essential, in the monistic conception of evolution, to distinguish between the laws of conservative and progressive heredity. Conservative heredity maintains from generation to generation the enduring characters of the species. Each organism transmits to its descendants a part of the morphological and physiological qualities that it has received from its parents and ancestors. On the other hand, progressive heredity brings new characters to the species--characters that were not found in preceding generations. Each organism may transmit to its offspring a part of the morphological and physiological features that it has itself acquired, by adaptation, in the course of its individual career, through the use or disuse of particular organs, the influence of environment, climate, nutrition, etc. At that time I gave the name of "progressive heredity" to this inheritance of acquired characters, as a short and convenient expression, but have since changed the term to "transformative heredity" (as distinguished from conservative). This term is preferable, as inherited regressive modifications (degeneration, retrograde metamorphosis, etc.) come under the same head.

Transformative heredity--or the transmission of acquired characters--is one of the most important principles in evolutionary science. Unless we admit it most of the facts of comparative anatomy and physiology are inexplicable. That was the conviction of Darwin no less than of Lamarck, of Spencer as well as Virchow, of Huxley as well as Gegenbaur, indeed of the great majority of speculative biologists. This fundamental principle was for the first time called in question and assailed in 1885 by August Weismann of Freiburg, the eminent zoologist to whom the theory of evolution owes a great deal of valuable support, and who has attained distinction by his extension of the theory of selection. In explanation of the phenomena of heredity he introduced a new theory, the "theory of the continuity of the germ-plasm." According to him the living substance in all organisms consists of two quite distinct kinds of plasm, somatic and germinal. The permanent germ-plasm, or the active substance of the two germ-cells (egg-cell and sperm-cell), passes unchanged through a series of generations, and is not affected by environmental influences. The environment modifies only the soma-plasm, the organs and tissues of the body. The modifications that these parts undergo through the influence of the environment or their own activity (use and habit), do not affect the germ-plasm, and cannot therefore be transmitted.

This theory of the continuity of the germ-plasm has been expounded by Weismann during the last twenty-four years in a number of able volumes, and is regarded by many biologists, such as Mr. Francis Galton, Sir E. Ray Lankester, and Professor J. Arthur Thomson (who has recently made a thorough-going defence of it in his important work Heredity),[129] as the most striking advance in evolutionary science. On the other hand, the theory has been rejected by Herbert Spencer, Sir W. Turner, Gegenbaur, Kliker, Hertwig, and many others. For my part I have, with all respect for the distinguished Darwinian, contested the theory from the first, because its whole foundation seems to me erroneous, and its deductions do not seem to be in accord with the main facts of comparative morphology and physiology. Weismann's theory in its entirety is a finely conceived molecular hypothesis, but it is devoid of empirical basis. The notion of the absolute and permanent independence of the germ-plasm, as distinguished from the soma-plasm, is purely speculative; as is also the theory of germinal selection. The determinants, ids, and idants, are purely hypothetical elements. The experiments that have been devised to demonstrate their existence really prove nothing.

It seems to me quite improper to describe this hypothetical structure as "Neodarwinism." Darwin was just as convinced as Lamarck of the transmission of acquired characters and its great importance in the scheme of evolution. I had the good fortune to visit Darwin at Down three times and discuss with him the main principles of his system, and on each occasion we were fully agreed as to the incalculable importance of what I may call transformative inheritance. It is only proper to point out that Weismann's theory of the germ-plasm is in express contradiction to the fundamental principles of Darwin and Lamarck. Nor is it more acceptable in what one may call its "ultradarwinism"--the idea that the theory of selection explains everything in the evolution of the organic world. This belief in the "omnipotence of natural selection" was not shared by Darwin himself. Assuredly, I regard it as of the utmost value, as the process of natural selection through the struggle for life affords an explanation of the mechanical origin of the adapted organisation. It solves the great problem: how could the finely adapted structure of the animal or plant body be formed unless it was built on a preconceived plan? It thus enables us to dispense with the teleology of the metaphysician and the dualist, and to set aside the old mythological and poetic legends of creation. The idea had occurred in vague form to the great Empedocles 2000 years before the time of Darwin, but it was reserved for modern research to give it ample expression. Nevertheless, natural selection does not of itself give the solution of all our evolutionary problems. It has to be taken in conjunction with the transformism of Lamarck, with which it is in complete harmony.

The monumental greatness of Charles Darwin, who surpasses every other student of science in the nineteenth century by the loftiness of his monistic conception of nature and the progressive influence of his ideas, is perhaps best seen in the fact that not one of his many successors has succeeded in modifying his theory of descent in any essential point or in discovering an entirely new standpoint in the interpretation of the organic world. Neither Neli nor Weismann, neither De Vries nor Roux, has done this. Neli, in his Mechanisch-Physiologische Theorie der Abstammungslehre[130] which is to a great extent in agreement with Weismann, constructed a theory of the idioplasm, that represents it (like the germ-plasm) as developing continuously in a definite direction from internal causes. But his internal "principle of progress" is at the bottom just as teleological as the vital force of the Vitalists, and the micella structure of the idioplasm is just as hypothetical as the "dominant" structure of the germ-plasm. In 1889 Moritz Wagner sought to explain the origin of species by migration and isolation, and on that basis constructed a special "migration-theory." This, however, is not out of harmony with the theory of selection. It merely elevates one single factor in the theory to a predominant position. Isolation is only a special case of selection, as I had pointed out in the fifteenth chapter of my Natural history of creation. The "mutation-theory" of De Vries,[131] that would explain the origin of species by sudden and saltatory variations rather than by gradual modification, is regarded by many botanists as a great step in advance, but it is generally rejected by zoologists. It affords no explanation of the facts of adaptation, and has no causal value.

Much more important than these theories is that of Wilhelm Roux[132] of "the struggle of parts within the organism, a supplementation of the theory of mechanical adaptation." He explains the functional autoformation of the purposive structure by a combination of Darwin's principle of selection with Lamarck's idea of transformative heredity, and applies the two in conjunction to the facts of histology. He lays stress on the significance of functional adaptation, which I had described in 1866, under the head of cumulative adaptation, as the most important factor in evolution. Pointing out its influence in the cell-life of the tissues, he puts "cellular selection" above "personal selection," and shows how the finest conceivable adaptations in the structure of the tissue may be brought about quite mechanically, without preconceived plan. This "mechanical teleology" is a valuable extension of Darwin's monistic principle of selection to the whole field of cellular physiology and histology, and is wholly destructive of dualistic vitalism.

The most important advance that evolution has made since Darwin and the most valuable amplification of his theory of selection is, in my opinion, the work of Richard Semon: Die Mneme als erhaltendes Prinzip im Wechsel des organischen Geschehens.[133] He offers a psychological explanation of the facts of heredity by reducing them to a process of (unconscious) memory. The physiologist Ewald Hering had shown in 1870 that memory must be regarded as a general function of organic matter, and that we are quite unable to explain the chief vital phenomena, especially those of reproduction and inheritance, unless we admit this unconscious memory. In my essay Die Perigenesis der Plastidule[134] I elaborated this far-reaching idea, and applied the physical principle of transmitted motion to the plastidules, or active molecules of plasm. I concluded that "heredity is the memory of the plastidules, and variability their power of comprehension." This "provisional attempt to give a mechanical explanation of the elementary processes of evolution" I afterwards extended by showing that sensitiveness is (as Carl

N 鋄 eli, Ernst Mach, and Albrecht Rau express it) a general quality of matter. This form of panpsychism finds its simplest expression in the "trinity of substance."

To the two fundamental attributes that Spinoza ascribed to substance-- Extension (matter as occupying space) and Cogitation (energy, force)--we now add the third fundamental quality of Psychoma (sensitiveness, soul). I further elaborated this trinitarian conception of substance in the nineteenth chapter of my Die Lebenswunder (1904),[135] and it seems to me well calculated to afford a monistic solution of many of the antitheses of philosophy.

This important Mneme-theory of Semon and the luminous physiological experiments and observations associated with it not only throw considerable light on transformative inheritance, but provide a sound physiological foundation for the biogenetic law. I had endeavoured to show in 1874, in the first chapter of my Anthropogenie,[136] that this fundamental law of organic evolution holds good generally, and that there is everywhere a direct causal connection between ontogeny and phylogeny. "Phylogenesis is the mechanical cause of ontogenesis;" in other words, "The evolution of the stem or race is--in accordance with the laws of heredity and adaptation--the real cause of all the changes that appear, in a condensed form, in the development of the individual organism from the ovum, in either the embryo or the larva."

It is now fifty years since Charles Darwin pointed out, in the thirteenth chapter of his epoch-making Origin of Species, the fundamental importance of embryology in connection with his theory of descent:

"The leading facts in embryology, which are second to none in importance, are explained on the principle of variations in the many descendants from some one ancient progenitor, having appeared at a not very early period of life, and having been inherited at a corresponding period."[137]

He then shows that the striking resemblance of the embryos and larvae of closely related animals, which in the mature stage belong to widely different species and genera, can only be explained by their descent from a common progenitor. Fritz Muller made a closer study of these important phenomena in the instructive instance of the Crustacean larva, as given in his able work Darwin[138] (1864). I then, in 1872, extended the range so as to include all animals (with the exception of the unicellular Protozoa) and showed, by means of the theory of the Gastraea, that all multicellular, tissue-forming animals--all the Metazoa--develop in essentially the same way from the primary germ-layers.

I conceived the embryonic form, in which the whole structure consists of only two layers of cells, and is known as the gastrula, to be the ontogenetic recapitulation, maintained by tenacious heredity, of a primitive common progenitor of all the Metazoa, the Gastraea. At a later date (1895) Monticelli discovered that this conjectural ancestral form is still preserved in certain primitive Coelenterata--Pemmatodiscus, Kunstleria, and the nearly-related Orthonectida.

The general application of the biogenetic law to all classes of animals and plants has been proved in my Systematische Phylogenie.[139] It has, however, been frequently challenged, both by botanists and zoologists, chiefly owing to the fact that many have failed to distinguish its two essential elements, palingenesis and cenogenesis. As early as 1874 I had emphasised, in the first chapter of my Evolution of Man, the importance of discriminating carefully between these two sets of phenomena:

"In the evolutionary appreciation of the facts of embryology we must take particular care to distinguish sharply and clearly between the primary, palingenetic evolutionary processes and the secondary, cenogenetic processes. The palingenetic phenomena, or embryonic recapitulations, are due to heredity, to the transmission of characters from one generation to another. They enable us to draw direct inferences in regard to corresponding structures in the development of the species (e.g. the chorda or the branchial arches in all vertebrate embryos). The cenogenetic phenomena, on the other hand, or the embryonic variations, cannot be traced to inheritance from a mature ancestor, but are due to the adaption of the embryo or the larva to certain conditions of its individual development (e.g. the amnion, the allantois, and the vitelline arteries in the embryos of the higher vertebrates). These cenogenetic phenomena are later additions; we must not infer from them that there were corresponding processes in the ancestral history, and hence they are apt to mislead."

The fundamental importance of these facts of comparative anatomy, atavism, and the rudimentary organs, was pointed out by Darwin in the first part of his classic work, The Descent of Man and Selection in Relation to Sex (1871).[140] In the "General summary and conclusion" (chap. xxi.) he was able to say, with perfect justice: "He who is not content to look, like a savage, at the phenomena of nature as disconnected, cannot any longer believe that man is the work of a separate act of creation. He will be forced to admit that the close resemblance of the embryo of man to that, for instance, of a dog-- the construction of his skull, limbs, and whole frame on the same plan with that of other mammals, independently of the uses to which the parts may be put--the occasional reappearance of various structures, for instance of several muscles, which man does not normally possess, but which are common to the Quadrumana--and a crowd of analogous facts--all point in the plainest manner to the conclusion that man is the co-descendant with other mammals of a common progenitor."

These few lines of Darwin's have a greater scientific value than hundreds of those so-called "anthropological treatises," which give detailed descriptions of single organs, or mathematical tables with series of numbers and what are claimed to be "exact analyses," but are devoid of synoptic conclusions and a philosophical spirit.

Charles Darwin is not generally recognised as a great anthropologist, nor does the school of modern anthropologists regard him as a leading authority. In Germany, especially, the great majority of the members of the anthropological societies took up an attitude of hostility to him from the very beginning of the controversy in 1860. The Descent of Man was not merely rejected, but even the discussion of it was forbidden on the ground that it was "unscientific."

The centre of this inveterate hostility for thirty years--especially after 1877-- was Rudolph Virchow of Berlin, the leading investigator in pathological anatomy, who did so much for the reform of medicine by his establishment of cellular pathology in 1858. As a prominent representative of "exact" or "descriptive" anthropology, and lacking a broad equipment in comparative anatomy and ontogeny, he was unable to accept the theory of descent. In earlier years, and especially during his splendid period of activity at Walzburg

(1848-1856), he had been a consistent free-thinker, and had in a number of able articles (collected in his Gesammelte Abhandlungen)[141] upheld the unity of human nature, the inseparability of body and spirit. In later years at Berlin, where he was more occupied with political work and sociology (especially after 1866), he abandoned the positive monistic position for one of agnosticism and scepticism, and made concessions to the dualistic dogma of a spiritual world apart from the material frame.

In the course of a Scientific Congress at Munich in 1877 the conflict of these antithetic views of nature came into sharp relief. At this memorable Congress I had undertaken to deliver the first address (September 18th) on the subject of "Modern evolution in relation to the whole of science." I maintained that Darwin's theory not only solved the great problem of the origin of species, but that its implications, especially in regard to the nature of man, threw considerable light on the whole of science, and on anthropology in particular. The discovery of the real origin of man by evolution from a long series of mammal ancestors threw light on his place in nature in every respect, as Huxley had already shown in his excellent lectures of 1863. Just as all the organs and tissues of the human body had originated from those of the nearest related mammals, certain ape-like forms, so we were bound to conclude that his mental qualities also had been derived from those of his extinct primate ancestor.

This monistic view of the origin and nature of man, which is now admitted by nearly all who have the requisite acquaintance with biology, and approach the subject without prejudice, encountered a sharp opposition at that time. The opposition found its strongest expression in an address that Virchow delivered at Munich four days afterwards (September 22nd), on "The freedom of science in the modern State." He spoke of the theory of evolution as an unproved hypothesis, and declared that it ought not to be taught in the schools, because it was dangerous to the State. "We must not," he said, "teach that man has descended from the ape or any other animal." When Darwin, usually so lenient in his judgment, read the English translation of Virchow's speech, he expressed his disapproval in strong terms. But the great authority that Virchow had--an authority well founded in pathology and sociology--and his prestige as president of the German Anthropological Society, had the effect of preventing any member of the Society from raising serious opposition to him for thirty years. Numbers of journals and treatises repeated his dogmatic statement: "It is quite certain that man has descended neither from the ape nor from any other animal." In this he persisted till his death in 1902. Since that time the whole position of German anthropology has changed. The question is no longer whether man was created by a distinct supernatural act or evolved from other mammals, but to which line of the animal hierarchy we must look for the actual series of ancestors. The interested reader will find an account of this "battle of Munich" (1877) in my three Berlin lectures (April, 1905), Der Kampf um die Entwickelungs-Gedanken.[142]

The main points in our genealogical tree were clearly recognised by Darwin in the sixth chapter of the Descent of Man. Lowly organised fishes, like the lancelot (Amphioxus), are descended from lower invertebrates resembling the larvae of an existing Tunicate (Appendicularia). From these primitive fishes were evolved higher fishes of the ganoid type and others of the type of Lepidosiren (Dipneusta). It is a very small step from these to the Amphibia:

"In the class of animals the steps are not difficult to conceive which led from the ancient Monotremata to the ancient Marsupials; and from these to the early progenitors of the placental mammals. We may thus ascend to the Lemuridae; and the interval is not very wide from these to the Simiadae. The Simiadae then branched off into two great stems, the New World and Old World monkeys; and from the latter, at a remote period, Man, the wonder and glory of the Universe, proceeded."[143]

In these few lines Darwin clearly indicated the way in which we were to conceive our ancestral series within the vertebrates. It is fully confirmed by all the arguments of comparative anatomy and embryology, of palaeontology and physiology; and all the research of the subsequent forty years have gone to establish it. The deep interest in geology which Darwin maintained throughout his life and his complete knowledge of palaeontology enabled him to grasp the fundamental importance of the palaeontological record more clearly than anthropologists and zoologists usually do.

There has been much debate in subsequent decades whether Darwin himself maintained that man was descended from the ape, and many writers have sought to deny it. But the lines I have quoted verbatim from the conclusion of the sixth chapter of the Descent of Man (1871) leave no doubt that he was as firmly convinced of it as was his great precursor Jean Lamarck in 1809. Moreover, Darwin adds, with particular explicitness, in the "general summary and conclusion" (chap. xxi.) of that standard work:[144]

"By considering the embryological structure of man--the homologies which he presents with the lower animals,--the rudiments which he retains,--and the reversions to which he is liable, we can partly recall in imagination the former condition of our early progenitors; and can approximately place them in their proper place in the zoological series. We thus learn that man is descended from a hairy, tailed quadruped, probably arboreal in its habits, and an inhabitant of the Old World. This creature, if its whole structure had been examined by a naturalist, would have been classed amongst the Quadrumana, as surely as the still more ancient progenitor of the Old and New World monkeys."

These clear and definite lines leave no doubt that Darwin--so critical and cautious in regard to important conclusions--was quite as firmly convinced of the descent of man from the apes (the Catarrhinae, in particular) as Lamarck was in 1809 and Huxley in 1863.

It is to be noted particularly that, in these and other observations on the subject, Darwin decidedly assumes the monophyletic origin of the mammals, including man. It is my own conviction that this is of the greatest importance. A number of difficult questions in regard to the development of man, in respect of anatomy, physiology, psychology, and embryology, are easily settled if we do not merely extend our progonotaxis to our nearest relatives, the anthropoid apes and the tailed monkeys from which these have descended, but go further back and find an ancestor in the group of the Lemuridae, and still further back to the Marsupials and Monotremata. The essential identity of all the Mammals in point of anatomical structure and embryonic development--in spite of their astonishing differences in external appearance and habits of life--is so palpably significant that modern zoologists are agreed in the hypothesis that they have all sprung from a common root, and that this root may be sought in the earlier Palaeozoic Amphibia.

The fundamental importance of this comparative morphology of the Mammals, as a sound basis of scientific anthropology, was recognised just before the beginning of the nineteenth century, when Lamarck first emphasised (1794) the division of the animal kingdom into Vertebrates and Invertebrates. Even thirteen years earlier (1781), when Goethe made a close study of the mammal skeleton in the Anatomical Institute at Jena, he was intensely interested to find that the composition of the skull was the same in man as in the other mammals. His discovery of the os inter-maxillare in man (1784), which was contradicted by most of the anatomists of the time, and his ingenious "vertebral theory of the skull," were the splendid fruit of his morphological studies. They remind us how Germany's greatest philosopher and poet was for many years ardently absorbed in the comparative anatomy of man and the mammals, and how he divined that their wonderful identity in structure was no mere superficial resemblance, but pointed to a deep internal connection. In my Generelle Morphologie (1866), in which I published the first attempts to construct phylogenetic trees, I have given a number of remarkable theses of Goethe, which may be called "phyletic prophecies." They justify us in regarding him as a precursor of Darwin.

In the ensuing forty years I have made many conscientious efforts to penetrate further along that line of anthropological research that was opened up by Goethe, Lamarck, and Darwin. I have brought together the many valuable results that have constantly been reached in comparative anatomy, physiology, ontogeny, and palaeontology, and maintained the effort to reform the classification of animals and plants in an evolutionary sense. The first rough drafts of pedigrees that were published in the Generelle Morphologie have been improved time after time in the ten editions of my Natlich fungsgeschichte (1868-1902).[145] A sounded basis for my phyletic hypotheses, derived from a discriminating combination of the three great records--morphology, ontogeny, and palaeontology--was provided in the three volumes of my Systematische Phylogenie[146] (1894 Protists and Plants, 1895 Vertebrates, 1896 Invertebrates).

In my Anthropogenie[147] I endeavoured to employ all the known facts of comparative ontogeny (embryology) for the purpose of completing my scheme of human phylogeny (evolution). I attempted to sketch the historical development of each organ of the body, beginning with the most elementary structures in the germ-layers of the Gastraea. At the same time I drew up a corrected statement of the most important steps in the line of our ancestral series.

At the fourth International Congress of Zoology at Cambridge (August 26th, 1898) I delivered an address on "Our present knowledge of the Descent of Man." It was translated into English, enriched with many valuable notes and additions, by my friend and pupil in earlier days Dr. Hans Gadow (Cambridge), and published under the title: The Last Link: our present knowledge of the Descent of Man[148] The determination of the chief animal forms that occur in the line of our ancestry is there restricted to thirty types, and these are distributed in six main groups.

The first half of this "Progonotaxis hominis," which has no support from fossil evidence, comprises three groups: (i) Protista (unicellular organisms, 1-5): (ii) Invertebrate Metazoa (Coelenteria 6-8, Vermalia 9-11): (iii) Monorrhine Vertebrates (Acrania 12-13, Cyclostoma 14-15). The second half, which is based on fossil records, also comprises three groups: (iv) Palaeozoic cold-blooded Craniota (Fishes 16-18, Amphibia 19, Reptiles 20): (v) Mesozoic Mammals (Monotrema 21, Marsupialia 22, Mallotheria 23): (vi) Cenozoic Primates (Lemuridae 24-25, Tailed Apes 26-27, Anthropomorpha 28-30). An improved and enlarged edition of this hypothetic "Progonotaxis hominis" was published in 1908, in my essay Unsere Ahnenreihe.[149]

If I have succeeded in furthering, in some degree, by these anthropological works, the solution of the great problem of Man's place in nature, and particularly in helping to trace the definite stages in our ancestral series, I owe the success, not merely to the vast progress that biology has made in the last half century, but largely to the luminous example of the great investigators who have applied themselves to the problem, with so much assiduity and genius, for a century and a quarter--I mean Goethe and Lamarck, Gegenbaur and Huxley, but, above all, Charles Darwin. It was the great genius of Darwin that first brought together that symmetrical temple of scientific knowledge, the theory of descent. It was Darwin who put the crown on the edifice by his theory of natural selection. Not until this broad inductive law was firmly established was it possible to vindicate the special conclusion, the descent of man from a series of other Vertebrates. By his illuminating discovery Darwin did more for anthropology than thousands of those writers, who are more specifically titled anthropologists, have done by their technical treatises. We may, indeed, say that it is not merely as an exact observer and ingenious experimenter, but as a distinguished anthropologist and far-seeing thinker, that Darwin takes his place among the greatest men of science of the nineteenth century.

To appreciate fully the immortal merit of Darwin in connection with anthropology, we must remember that not only did his chief work, The Origin of Species, which opened up a new era in natural history in 1859, sustain the most virulent and widespread opposition for a lengthy period, but even thirty years later, when its principles were generally recognised and adopted, the application of them to man was energetically contested by many high scientific authorities. Even Alfred Russel Wallace, who discovered the principle of natural selection independently in 1858, did not concede that it was applicable to the higher mental and moral qualities of man. Dr. Wallace still holds a spiritualist and dualist view of the nature of man, contending that he is composed of a material frame (descended from the apes) and an immortal immaterial soul (infused by a higher power). This dual conception, moreover, is still predominant in the wide circles of modern theology and metaphysics, and has the general and influential adherence of the more conservative classes of society.

In strict contradiction to this mystical dualism, which is generally connected with teleology and vitalism, Darwin always maintained the complete unity of human nature, and showed convincingly that the psychological side of man was developed, in the same way as the body, from the less advanced soul of the anthropoid ape, and, at a still more remote period, from the cerebral functions of the older vertebrates. The eighth chapter of the Origin of Species, which is devoted to instinct, contains weighty evidence that the instincts of animals are subject, like all other vital processes, to the general laws of historic development. The special instincts of particular species were formed by adaptation, and the modifications thus acquired were handed on to posterity by heredity; in their formation and preservation natural selection plays the same part as in the transformation of every other physiological function. The higher moral qualities of civilised man have been derived from the lower mental functions of the uncultivated barbarians and savages, and these in turn from the social instincts of the mammals. This natural and monistic psychology of Darwin's was afterwards more fully developed by his friend George Romanes in his excellent works Mental Evolution in Animals and Mental Evolution in Man.[150]

Many valuable and most interesting contributions to this monistic psychology of man were made by Darwin in his fine work on The Descent of Man and Selection in Relation to Sex, and again in his supplementary work, The Expression of the Emotions in Man and Animals. To understand the historical development of Darwin's anthropology one must read his life and the introduction to The Descent of Man. From the moment that he was convinced of the truth of the principle of descent--that is to say, from his thirtieth year, in 1838--he recognised clearly that man could not be excluded from its range. He recognised as a logical necessity the important conclusion that "man is the co-descendant with other species of some ancient, lower, and extinct form." For many years he gathered notes and arguments in support of this thesis, and for the purpose of showing the probable line of man's ancestry. But in the first edition of The Origin of Species (1859) he restricted himself to the single line, that by this work "light would be thrown on the origin of man and his history." In the fifty years that have elapsed since that time the science of the origin and nature of man has made astonishing progress, and we are now fairly agreed in a monistic conception of nature that regards the whole universe, including man, as a wonderful unity, governed by unalterable and eternal laws. In my philosophical book Die Weltresel (1899)[151] and in the supplementary volume Die Lebenswunder (1904)[152] I have endeavoured to show that this pure monism is securely established, and that the admission of the all-powerful rule of the same principle of evolution throughout the universe compels us to formulate a single supreme law--the all-embracing "Law of Substance," or the united laws of the constancy of matter and the conservation of energy. We should never have reached this supreme general conception if Charles Darwin--a "monistic philosopher" in the true sense of the word--had not prepared the way by his theory of descent by natural selection, and crowned the great work of his life by the association of this theory with a naturalistic anthropology.

FOOTNOTES:

[Footnote 127: Generelle Morphologie der Organismen, 2 vols., Berlin, 1866.]

[Footnote 128: Eng. transl.; The History of Creation, London, 1876.]

[Footnote 129: London, 1908.]

[Footnote 130: Munich, 1884.]

[Footnote 131: Die Mutationstheorie, Leipzig, 1903.]

[Footnote 132: Der Kampf der Theile im Organismus, Leipzig, 1881.]

[Footnote 133: Leipzig, 1904.]

[Footnote 134: Berlin, 1876.]

[Footnote 135: Wonders of Life, London and New York, 1904.]

[Footnote 136: Eng. transl.; The Evolution of Man, 2 vols., London, 1879 and 1905.]

[Footnote 137: Origin of Species (6th edit.), p. 396.]

[Footnote 138: Eng. transl.; Facts and Arguments for Darwin, London, 1869.]

[Footnote 139: 3 vols., Berlin, 1894-96.]

[Footnote 140: Descent of Man (Popular Edit.), p. 927.]

[Footnote 141: Gesammelte Abhandlungen zur wissenschaftlichen Medizin, Berlin, 1856.]

[Footnote 142: Eng. transl.; Last Words on Evolution, London, 1906.]

[Footnote 143: Descent of Man, (Popular Edit.), p. 255.]

[Footnote 144: Descent of Man, p. 930.]

[Footnote 145: Eng. transl.; The History of Creation, London, 1876.]

[Footnote 146: Berlin, 1894-96.]

[Footnote 147: Leipzig, 1874, 5th edit. 1905. Eng. transl.; The Evolution of Man, London, 1905.]

[Footnote 148: London, 1898.]

[Footnote 149: Festschrift zur 350-j鋒rigen Jubelfeier der Th鼈inger Universit鏗 Jena. Jena. 1908.]

[Footnote 150: London, 1885; 1888.]

[Footnote 151: The Riddle of the Universe, London and New York, 1900.]

[Footnote 152: The Wonders of Life, London and New York, 1904.]

VI

MENTAL FACTORS IN EVOLUTION

BY C. LLOYD MORGAN, LLD., F.R.S

In developing his conception of organic evolution Charles Darwin was of necessity brought into contact with some of the problems of mental evolution. In The Origin of Species he devoted a chapter to "the diversities of instinct and of the other mental faculties in animals of the same class."[153] When he passed to the detailed consideration of The Descent of Man, it was part of his object to show "that there is no fundamental difference between man and the higher mammals in their mental faculties."[154] "If no organic being excepting man," he said, "had possessed any mental power, or if his powers had been of a wholly different nature, from those of the lower animals, then we should never have been able to convince ourselves that our high faculties had been gradually developed."[155] In his discussion of The Expression of the Emotions it was important for his purpose "fully to recognise that actions readily become associated with other actions and with various states of the mind."[156] His hypothesis of sexual selection is largely dependent upon the exercise of choice on the part of the female and her preference for "not only the more attractive but at the same time the more vigourous and vicious males."[157] Mental processes and physiological processes were for Darwin closely correlated; and he accepted the conclusion "that the nervous system not only regulates most of the existing functions of the body, but has indirectly influenced the progressive development of various bodily structures and of certain mental qualities."[158]

Throughout his treatment, mental evolution was for Darwin incidental to and contributory to organic evolution. For specialised research in comparative and genetic psychology, as an independent field of investigation, he had neither the time nor the requisite training. None the less his writings and the spirit of his work have exercised a profound influence on this department of evolutionary thought. And, for those who follow Darwin's lead, mental evolution is still in a measure subservient to organic evolution. Mental processes are the accompaniments or concomitants of the functional activity of specially differentiated parts of the organism. They are in some way dependent on physiological and physical conditions. But though they are not physical in their nature, and though it is difficult or impossible to conceive that they are physical in their origin, they are, for Darwin and his followers, factors in the evolutionary process in its physical or organic aspect. By the physiologist within his special and well-defined universe of discourse they may be properly regarded as epiphenomena; but by the naturalist in his more catholic survey of nature they cannot be so regarded, and were not so regarded by Darwin. Intelligence has contributed to evolution of which it is in a sense a product.

The facts of observation or of inference which Darwin accepted are these: Conscious experience accompanies some of the modes of animal behaviour; it is concomitant with certain physiological processes; these processes are the outcome of development in the individual and evolution in the race; the accompanying mental processes undergo a like development. Into the subtle philosophical questions which arise out of the naive acceptance of such a creed it was not Darwin's province to enter; "I have nothing to do," he said,[159] "with the origin of the mental powers, any more than I have with that of life itself." He dealt with the natural history of organisms, including not only their structure but their modes of behaviour; with the natural history of the states of consciousness which accompany some of their actions; and with the relation of behaviour to experience. We will endeavour to follow Darwin in his modesty and candour in making no pretence to give ultimate explanations. But we must note one of the implications of this self-denying ordinance of science. Development and evolution imply continuity. For Darwin and his followers the continuity is organic through physical heredity. Apart from speculative hypothesis, legitimate enough in its proper place but here out of court, we know nothing of continuity of mental evolution as such: consciousness appears afresh in each succeeding generation. Hence it is that for those who follow Darwin's lead, mental evolution is and must ever be, within his universe of discourse, subservient to organic evolution. Only in so far as conscious experience, or its neural correlate, effects some changes in organic structure can it influence the course of heredity; and conversely only in so far as changes in organic structure are transmitted through heredity, is mental evolution rendered possible. Such is the logical outcome of Darwin's teaching.

Those who abide by the cardinal results of this teaching are bound to regard all behaviour as the expression of the functional activities of the living tissues of the organism, and all conscious experience as correlated with such activities. For the purposes of scientific treatment, mental processes are one mode of expression of the same changes of which the physiological processes accompanying behaviour are another mode of expression. This is simply accepted as a fact which others may seek to explain. The behaviour itself is the adaptive application of the energies of the organism; it is called forth by some form of presentation or stimulation brought to bear on the organism by the environment. This presentation is always an individual or personal matter. But in order that the organism may be fitted to respond to the presentation of the environment it must have undergone in some way a suitable preparation. According to the theory of evolution this preparation is primarily racial and is transmitted through heredity. Darwin's main thesis was that the method of preparation is predominantly by natural selection. Subordinate to racial preparation, and always dependent thereon, is individual or personal preparation through some kind of acquisition; of which the guidance of behaviour through individually won experience is a typical example. We here introduce the mental factor because the facts seem to justify the inference. Thus there are some modes of behaviour which are wholly and solely dependent upon inherited racial preparation; there are other modes of behaviour which are also dependent, in part at least, on individual preparation. In the former case the behaviour is adaptive on the first occurrence of the appropriate presentation; in the latter case accommodation to circumstances is only reached after a greater or less amount of acquired organic modification of structure, often accompanied (as we assume) in the higher animals by acquired experience. Logically and biologically the two classes of behaviour are clearly distinguishable: but the analysis of complex cases of behaviour where the two factors cooperate, is difficult and requires careful and critical study of life-history.

The foundations of the mental life are laid in the conscious experience that accompanies those modes of behaviour, dependent entirely on racial preparation, which may broadly be described as instinctive. In the eighth chapter of The Origin of Species Darwin says,[160] "I will not attempt any definition of instinct.... Every one understands what is meant, when it is said that instinct impels the cuckoo to migrate and to lay her eggs in other birds' nests. An action, which we ourselves require experience to enable us to perform, when performed by an animal, more especially by a very young one, without experience, and when performed by many individuals in the same way, without their knowing for what purpose it is performed, is usually said to be instinctive." And in the summary at the close of the chapter he says,[161] "I have endeavoured briefly to show that the mental qualities of our domestic animals vary, and that the variations are inherited. Still more briefly I have attempted to show that instincts vary slightly in a state of nature. No one will dispute that instincts are of the highest importance to each animal. Therefore there is no real difficulty, under changing conditions of life, in natural selection accumulating to any extent slight modifications of instinct which are in any way useful. In many cases habit or use and disuse have probably come into play."

Into the details of Darwin's treatment there is neither space nor need to enter. There are some ambiguous passages; but it may be said that for him, as for his followers to-day, instinctive behaviour is wholly the result of racial preparation transmitted through organic heredity. For the performance of the instinctive act no individual preparation under the guidance of personal experience is necessary. It is true that Darwin quotes with approval Huber's saying that "a little dose of judgment or reason often comes into play, even with animals low in the scale of nature."[162] But we may fairly interpret his meaning to be that in behaviour, which is commonly called instinctive, some element of Intelligent guidance is often combined. If this be conceded the strictly instinctive performance (or part of the performance) is the outcome of heredity and due to the direct transmission of parental or ancestral aptitudes. Hence the instinctive response as such depends entirely on how the nervous mechanism has been built up through heredity; while intelligent behaviour, or the intelligent factor in behaviour, depends also on how the nervous mechanism has been modified and moulded by use during its development and concurrently with the growth of individual experience in the customary situations of daily life. Of course it is essential to the Darwinian thesis that what Sir E. Ray Lankester has termed "educability," not less than instinct, is hereditary. But it is also essential to the understanding of this thesis that the differentiae of the hereditary factor should be clearly grasped.

For Darwin there were two modes of racial preparation, (1) natural selection, and (2) the establishment of individually acquired habit. He showed that instincts are subject to hereditary variation; he saw that instincts are also subject to modification through acquisition in the course of individual life. He believed that not only the variations but also, to some extent, the modifications are inherited. He therefore held that some instincts (the greater number) are due to natural selection but that others (less numerous) are due, or partly due, to the inheritance of acquired habits. The latter involve Lamarckian inheritance, which of late years has been the centre of so much controversy. It is noteworthy however that Darwin laid especial emphasis on the fact that many of the most typical and also the most complex instincts--those of neuter insects--do not admit of such an interpretation. "I am surprised," he says,[163] "that no one has hitherto advanced this demonstrative case of neuter insects, against the well-known doctrine of inherited habit, as advanced by Lamarck." None the less Darwin admitted this doctrine as supplementary to that which was more distinctively his own--for example in the case of the instincts of domesticated animals. Still, even in such cases, "it may be doubted," he says,[164] "whether any one would have thought of training a dog to point, had not some one dog naturally shown a tendency in this line ... so that habit and some degree of selection have probably concurred in civilising by inheritance our dogs." But in the interpretation of the instincts of domesticated animals, a more recently suggested hypothesis, that of organic selection,[165] may be helpful. According to this hypothesis any intelligent modification of behaviour which is subject to selection is probably coincident in direction with an inherited tendency to behave in this fashion. Hence in such behaviour there are two factors: (1) an incipient variation in the line of such behaviour, and (2) an acquired modification by which the behaviour is carried further along the same line. Under natural selection those organisms in which the two factors cooperate are likely to survive. Under artificial selection they are deliberately chosen out from among the rest.

Organic selection has been termed a compromise between the more strictly Darwinian and the Lamarckian principles of interpretation. But it is not in any sense a compromise. The principle of interpretation of that which is instinctive and hereditary is wholly Darwinian. It is true that some of the facts of observation relied upon by Lamarckians are introduced. For Lamarckians however the modifications which are admittedly factors in survival, are regarded as the parents of inherited variations; for believers in organic selection they are only the foster-parents or nurses. It is because organic selection is the direct outcome of and a natural extension of Darwin's cardinal thesis that some reference to it here is justifiable. The matter may be put with the utmost brevity as follows: (1) Variations (V) occur, some of which are in the direction of increased adaptation (+), others in the direction of decreased adaptation (-).

(2) Acquired modifications (M) also occur. Some of these are in the direction of increased accommodation to circumstances (+), while others are in the direction of diminished accommodation (-). Four major combinations are

(b) + V with - M, (c) - V with + M,

(a) + V with + M, (d) - V with - M.

Of these (d) must inevitably be eliminated while (a) are selected. The predominant survival of (a) entails the survival of the adaptive variations which are inherited. The contributory acquisitions (+ M) are not inherited; but there are none the less factors in determining the survival of the coincident variations. It is surely abundantly clear that this is Darwinism and has no tincture of Lamarck's essential principle, the inheritance of acquired characters.

Whether Darwin himself would have accepted this interpretation of some at least of the evidence put forward by Lamarckians is unfortunately a matter of conjecture. The fact remains that in his interpretation of instinct and in allied questions he accepted the inheritance of individually acquired modifications of behaviour and structure.

Darwin was chiefly concerned with instinct from the biological rather than from the psychological point of view. Indeed it must be confessed that, from the latter standpoint, his conception of instinct as a "mental faculty" which "impels" an animal to the performance of certain actions, scarcely affords a satisfactory basis for genetic treatment. To carry out the spirit of Darwin's teaching it is necessary to link more closely biological and psychological evolution. The first step towards this is to interpret the phenomena of instinctive behaviour in terms of stimulation and response. It may be well to take a particular case. Swimming on the part of a duckling is, from the biological point of view, a typical example of instinctive behaviour. Gently lower a recently hatched bird into water: coordinated movements of the limbs follow in rhythmical sequence. The behaviour is new to the individual though it is no doubt closely related to that of walking, which is no less instinctive. There is a group of stimuli afforded by the "presentation" which results from partial immersion: upon this there follows as a complex response an application of the functional activities in swimming; the sequence of adaptive application on the appropriate presentation is determined by racial preparation. We know, it is true, but little of the physiological details of what takes place in the central nervous system; but in broad outline the nature of the organic mechanism and the manner of its functioning may at least be provisionally conjectured in the present state of physiological knowledge. Similarly in the case of the pecking of newly-hatched chicks; there is a visual presentation, there is probably a cooperating group of stimuli from the alimentary tract in need of food, there is an adaptive application of the activities in a definite mode of behaviour. Like data are afforded in a great number of cases of instinctive procedure, sometimes occurring very early in life, not infrequently deferred until the organism is more fully developed, but all of them dependent upon racial preparation. No doubt there is some range of variation in the behaviour, just such variation as the theory of natural selection demands. But there can be no question that the higher animals inherit a bodily organisation and a nervous system, the functional working of which gives rise to those inherited modes of behaviour which are termed instinctive.

It is to be noted that the term "instinctive" is here employed in the adjectival form as a descriptive heading under which may be grouped many and various modes of behaviour due to racial preparation. We speak of these as inherited; but in strictness what is transmitted through heredity is the complex of anatomical and physiological conditions under which, in appropriate circumstances, the organism so behaves. So far the term "instinctive" has a restricted biological connotation in terms of behaviour. But the connecting link between biological evolution and psychological evolution is to be sought,--as Darwin fully realised,--in the phenomena of instinct, broadly considered. The term "instinctive" has also a psychological connotation. What is that connotation?

Let us take the case of the swimming duckling or the pecking chick, and fix our attention on the first instinctive performance. Grant that just as there is, strictly speaking, no inherited behaviour, but only the conditions which render such behaviour under appropriate circumstances possible; so too there is no inherited experience, but only the conditions which render such experience possible; then the cerebral conditions in both cases are the same. The biological behaviour-complex, including the total stimulation and the total response with the intervening or resultant processes in the sensorium, is accompanied by an experience-complex including the initial stimulation-consciousness and resulting response-consciousness. In the experience-complex are comprised data which in psychological analysis are grouped under the headings of cognition, affective tone and conation. But the complex is probably experienced as an unanalysed whole. If then we use the term "instinctive" so as to comprise all congenital modes of behaviour which contribute to experience, we are in a position to grasp the view that the net result in consciousness constitutes what we may term the primary tissue of experience. To the development of this experience each instinctive act contributes. The nature and manner of organisation of this primary tissue of experience are dependent on inherited biological aptitudes; but they are from the outset onwards subject to secondary development dependent on acquired aptitudes. Biological values are supplemented by psychological values in terms of satisfaction or the reverse.

In our study of instinct we have to select some particular phase of animal behaviour and isolate it so far as is possible from the life of which it is a part. But the animal is a going concern, restlessly active in many ways. Many instinctive performances, as Darwin pointed out,[166] are serial in their nature. But the whole of active life is a serial and coordinated business. The particular instinctive performance is only an episode in a life-history, and every mode of behaviour is more or less closely correlated with other modes. This coordination of behaviour is accompanied by a correlation of the modes of primary experience. We may classify the instinctive modes of behaviour and their accompanying modes of instinctive experience under as many heads as may be convenient for our purposes of interpretation, and label them instincts of self-preservation, of pugnacity, of acquisition, the reproductive instincts, the parental instincts, and so forth. An instinct, in this sense of the term (for example the parental instinct), may be described as a specialised part of the primary tissue of experience differentiated in relation to some definite biological end. Under such an instinct will fall a large number of particular and often well-defined modes of behaviour, each with its own peculiar mode of experience.

It is no doubt exceedingly difficult as a matter of observation and of inference securely based thereon to distinguish what is primary from what is in part due to secondary acquisition--a fact which Darwin fully appreciated. Animals are educable in different degrees; but where they are educable they begin to profit by experience from the first. Only, therefore, on the occasion of the first instinctive act of a given type can the experience gained be regarded as wholly primary; all subsequent performance is liable to be in some degree, sometimes more, sometimes less, modified by the acquired disposition which the initial behaviour engenders. But the early stages of acquisition are always along the lines predetermined by instinctive differentiation. It is the task of comparative psychology to distinguish the primary tissue of experience from its secondary and acquired modifications. We cannot follow up the matter in further detail. It must here suffice to suggest that this conception of instinct as a primary form of experience lends itself better to natural history treatment than Darwin's conception of an impelling force, and that it is in line with the main trend of Darwin's thought.

In a characteristic work,--characteristic in wealth of detail, in closeness and fidelity of observation, in breadth of outlook, in candour and modesty,-- Darwin dealt with The Expression of the Emotions in Man and Animals. Sir Charles Bell in his Anatomy of Expression had contended that many of man's facial muscles had been specially created for the sole purpose of being instrumental in the expression of his emotions. Darwin claimed that a natural explanation, consistent with the doctrine of evolution, could in many cases be given and would in other cases be afforded by an extension of the principles he advocated. "No doubt," he said,[167] "as long as man and all other animals are viewed as independent creations, an effectual stop is put to our natural desire to investigate as far as possible the causes of Expression. By this doctrine, anything and everything can be equally well explained.... With mankind, some expressions ... can hardly be understood, except on the belief that man once existed in a much lower and animal-like condition. The community of certain expressions in distinct though allied species ... is rendered somewhat more intelligible, if we believe in their descent from a common progenitor. He who admits on general grounds that the structure and habits of all animals have been gradually evolved, will look at the whole subject of Expression in a new and interesting light."

Darwin relied on three principles of explanation. "The first of these principles is, that movements which are serviceable in gratifying some desire, or in relieving some sensation, if often repeated, become so habitual that they are performed, whether or not of any service, whenever the same desire or sensation is felt, even in a very weak degree."[168] The modes of expression which fall under this head have become instinctive through the hereditary transmission of acquired habit. "As far as we can judge, only a few expressive movements are learnt by each individual; that is, were consciously and voluntarily performed during the early years of life for some definite object, or in imitation of others, and then became habitual. The far greater number of the movements of expression, and all the more important ones, are innate or inherited; and such cannot be said to depend on the will of the individual. Nevertheless, all those included under our first principle were at first voluntarily performed for a definite object,--namely, to escape some danger, to relieve some distress, or to gratify some desire."[169]

"Our second principle is that of antithesis. The habit of voluntarily performing opposite movements under opposite impulses has become firmly established in us by the practice of our whole lives. Hence, if certain actions have been regularly performed, in accordance with our first principle, under a certain frame of mind, there will be a strong and involuntary tendency to the performance of directly opposite actions, whether or not these are of any use, under the excitement of an opposite frame of mind."[170] This principle of antithesis has not been widely accepted. Nor is Darwin's own position easy to grasp.

"Our third principle," he says,[171] "is the direct action of the excited nervous system on the body, independently of the will, and independently, in large part, of habit. Experience shows that nerve-force is generated and set free whenever the cerebro-spinal system is excited. The direction which this nerve-force follows is necessarily determined by the lines of connection between the nerve-cells, with each other and with various parts of the body."

Lack of space prevents our following up the details of Darwin's treatment of expression. Whether we accept or do not accept his three principles of explanation we must regard his work as a masterpiece of descriptive analysis, packed full of observations possessing lasting value. For a further development of the subject it is essential that the instinctive factors in expression should be more fully distinguished from those which are individually acquired--a difficult task--and that the instinctive factors should be rediscussed in the light of modern doctrines of heredity, with a view to determining whether Lamarckian inheritance, on which Darwin so largely relied, is necessary for an interpretation of the facts.

The whole subject as Darwin realised is very complex. Even the term "expression" has a certain amount of ambiguity. When the emotion is in full flood, the animal fights, flees, or faints. Is this full-tide effect to be regarded as expression; or are we to restrict the term to the premonitory or residual effects--the bared canine when the fighting mood is being roused, the ruffled fur when reminiscent representations of the object inducing anger cross the mind? Broadly considered both should be included. The activity of premonitory expression as a means of communication was recognised by Darwin; he might, perhaps, have emphasised it more strongly in dealing with the lower animals. Man so largely relies on a special means of communication, that of language, that he sometimes fails to realise that for animals with their keen powers of perception, and dependent as they are on such means of communication, the more strictly biological means of expression are full of subtle suggestiveness. Many modes of expression, otherwise useless, are signs of behaviour that may be anticipated,--signs which stimulate the appropriate attitude of response. This would not, however, serve to account for the utility of the organic accompaniments--heart-affection, respiratory changes, vaso-motor effects and so forth, together with heightened muscular tone,--on all of which Darwin lays stress[172] under his third principle. The biological value of all this is, however, of great importance, though Darwin was hardly in a position to take it fully into account.

Having regard to the instinctive and hereditary factors of emotional expression we may ask whether Darwin's third principle does not alone suffice as an explanation. Whether we admit or reject Lamarckian inheritance it would appear that all hereditary expression must be due to pre-established connections within the central nervous system and to a transmitted provision for coordinated response under the appropriate stimulation. If this be so, Darwin's first and second principles are subordinate and ancillary to the third, an expression, so far as it is instinctive or heredity, being "the direct result of the constitution of the nervous system."

Darwin accepted the emotions themselves as hereditary or acquired states of mind and devoted his attention to their expression. But these emotions themselves are genetic products and as such dependent on organic conditions. It remained, therefore, for psychologists who accepted evolution and sought to build on biological foundations to trace the genesis of these modes of animal and human experience. The subject has been independently developed by Professors Lange and James;[173] and some modification of their view is regarded by many evolutionists as affording the best explanation of the facts. We must fix our attention on the lower emotions, such as anger or fear, and on their first occurrence in the life of the individual organism. It is a matter of observation that if a group of young birds which have been hatched in an incubator are frightened by an appropriate presentation, auditory or visual, they instinctively respond in special ways. If we speak of this response as the expression, we find that there are many factors. There are certain visible modes of behaviour, crouching at once, scattering and then crouching, remaining motionless, the braced muscles sustaining an attitude of arrest, and so forth, There are also certain visceral or organic effects, such as affections of the heart and respiration. These can be readily observed by taking the young bird in the hand. Other effects cannot be readily observed; vaso-motor changes, affections of the alimentary canal, the skin and so forth. Now the essence of the James-Lange view, as applied to these congenital effects, is that though we are justified in speaking of them as effects of the stimulation, we are not justified, without further evidence, in speaking of them as effects of the emotional state. May it not rather be that the emotion as a primary mode of experience is the concomitant of the net result of the organic situation--the initial presentation, the instinctive mode of behaviour, the visceral disturbances?

According to this interpretation the primary tissue of experience of the emotional order, felt as an unanalysed complex, is generated by the stimulation of the sensorium by afferent or incoming physiological impulses from the special senses, from the organs concerned in the responsive behaviour, from the viscere and vaso-motor system.

Some psychologists, however, contend that the emotional experience is generated in the sensorium prior to, and not subsequent to, the behaviour-response and the visceral disturbances. It is a direct and not an indirect outcome of the presentation to the special senses. Be this as it may, there is a growing tendency to bring into the closest possible relation, or even to identify, instinct and emotion in their primary genesis. The central core of all such interpretations is that instinctive behaviour and experience, its emotional accompaniments, and its expression, are but different aspects of the outcome of the same organic occurrences. Such emotions are, therefore, only a distinguishable aspect of the primary tissue of experience and exhibit a like differentiation. Here again a biological foundation is laid for a psychological doctrine of the mental development of the individual.

The intimate relation between emotion as a psychological mode of experience and expression as a group of organic conditions has an important bearing on biological interpretation. The emotion, as the psychological accompaniment of orderly disturbances in the central nervous system, profoundly influences behaviour and often renders it more vigourous and more effective. The utility of the emotions in the struggle for existence can, therefore, scarcely be over-estimated. Just as keenness of perception has survival-value; just as it is obviously subject to variation; just as it must be enhanced under natural selection, whether individually acquired increments are inherited or not; and just as its value lies not only in this or that special perceptive act but in its importance for life as a whole; so the vigourous effectiveness of activity has survival-value; it is subject to variation; it must be enhanced under natural selection; and its importance lies not only in particular modes of behaviour but in its value for life as a whole. If emotion and its expression as a congenital endowment are but different aspects of the same biological occurrence; and if this is a powerful supplement to vigour effectiveness and persistency of behaviour, it must on Darwin's principles be subject to natural selection.

If we include under the expression of the emotions not only the premonitory symptoms of the initial phases of the organic and mental state, not only the signs or conditions of half-tide emotion, but the full-tide manifestation of an emotion which dominates the situation, we are naturally led on to the consideration of many of the phenomena which are discussed under the head of sexual selection. The subject is difficult and complex, and it was treated by Darwin with all the strength he could summon to the task. It can only be dealt with here from a special point of view--that which may serve to illustrate the influence of certain mental factors on the course of evolution. From this point of view too much stress can scarcely be laid on the dominance of emotion during the period of courtship and pairing in the more highly organised animals. It is a period of maximum vigour, maximum activity, and, correlated with special modes of behaviour and special organic and visceral accompaniments, a period also of maximum emotional excitement. The combats of males, their dances and aerial evolutions, their elaborate behaviour and display, or the flood of song in birds, are emotional expressions which are at any rate coincident in time with sexual periodicity. From the combat of the males there follows on Darwin's principles the elimination of those which are deficient in bodily vigour, deficient in special structures, offensive or protective, which contribute to success, deficient in the emotional supplement of which persistent and whole-hearted fighting is the expression, and deficient in alertness and skill which are the outcome of the psychological development of the powers of perception. Few biologists question that we have here a mode of selection of much importance, though its influence on psychological evolution often fails to receive its due emphasis. Mr. Wallace[174] regards it as "a form of natural selection"; "to it," he says, "we must impute the development of the exceptional strength, size, and activity of the male, together with the possession of special offensive and defensive weapons, and of all other characters which arise from the development of these or are correlated with them." So far there is little disagreement among the followers of Darwin--for Mr. Wallace, with fine magnanimity, has always preferred to be ranked as such, notwithstanding his right, on which a smaller man would have constantly insisted, to the claim of independent originator of the doctrine of natural selection. So far with regard to sexual selection Darwin and Mr. Wallace are agreed; so far and no farther. For Darwin, says Mr. Wallace,[175] "has extended the principle into a totally different field of action, which has none of that character of constancy and of inevitable result that attaches to natural selection, including male rivalry; for by far the larger portion of the phenomena, which he endeavours to explain by the direct action of sexual selection, can only be so explained on the hypothesis that the immediate agency is female choice or preference. It is to this that he imputes the origin of all secondary sexual characters other than weapons of offence and defence.... In this extension of sexual selection to include the action of female choice or preference, and in the attempt to give to that choice such wide-reaching effects, I am unable to follow him more than a very little way."

Into the details of Mr. Wallace's criticisms it is impossible to enter here. We cannot discuss either the mode of origin of the variations in structure which have rendered secondary sexual characters possible or the modes of selection other than sexual which have rendered them, within narrow limits, specifically constant. Mendelism and mutation theories may have something to say on the subject when these theories have been more fully correlated with the basal principles of selection. It is noteworthy that Mr. Wallace says:[176] "Besides the acquisition of weapons by the male for the purpose of fighting with other males, there are some other sexual characters which may have been produced by natural selection. Such are the various sounds and odours which are peculiar to the male, and which serve as a call to the female or as an indication of his presence. These are evidently a valuable addition to the means of recognition of the two sexes, and are a further indication, that the pairing season has arrived; and the production, intensification, and differentiation of these sounds and odours are clearly within the power of natural selection. The same remark will apply to the peculiar calls of birds, and even to the singing of the males." Why the same remark should not apply to their colours and adornments is not obvious. What is obvious is that "means of recognition" and "indication that the pairing season has arrived" are dependent on the perceptive powers of the female who recognises and for whom the indication has meaning. The hypothesis of female preference, stripped of the aesthetic surplusage which is psychologically both unnecessary and unproven, is really only different in degree from that which Mr. Wallace admits in principle when he says that it is probable that the female is pleased or excited by the display.

Let us for our present purpose leave on one side and regard as sub judice the question whether the specific details of secondary sexual characters are the outcome of female choice. For us the question is whether certain psychological accompaniments of the pairing situation have influenced the course of evolution and whether these psychological accompaniments are themselves the outcome of evolution. As a matter of observation, specially differentiated modes of behaviour, often very elaborate, frequently requiring highly developed skill, and apparently highly charged with emotional tone, are the precursors of pairing. They are generally confined to the males, whose fierce combats during the period of sexual activity are part of the emotional manifestation. It is inconceivable that they have no biological meaning; and it is difficult to conceive that they have any other biological end than to evoke in the generally more passive female the pairing impulse. They, are based on instinctive foundations ingrained in the nervous constitution through natural (or may we not say sexual?) selection in virtue of their profound utility. They are called into play by a specialised presentation such as the sight or the scent of the female at, or a little in advance of, a critical period of the physiological rhythm. There is no necessity that the male should have any knowledge of the end to which his strenuous activity leads up. In presence of the female there is an elaborate application of all the energies of behaviour, just because ages of racial preparation have made him biologically and emotionally what he is--a functionally sexual male that must dance or sing or go through hereditary movements of display, when the appropriate stimulation comes. Of course after the first successful courtship his future behaviour will be in some degree modified by his previous experience. No doubt during his first courtship he is gaining the primary data of a peculiarly rich experience, instinctive and emotional. But the biological foundations of the behaviour of courtship are laid in the hereditary coordinations. It would seem that in some cases, not indeed in all, perhaps especially in those cases in which secondary sexual behaviour is most highly evolved,--correlative with the ardour of the male is a certain amount of reluctance in the female. The pairing act on her part only takes place after prolonged stimulation, for affording which the behaviour of male courtship is the requisite presentation. The most vigourous, defiant and mettlesome male is preferred just because he alone affords a contributory stimulation adequate to evoke the pairing impulse with its attendant emotional tone.

It is true that this places female preference or choice on a much lower psychological plane than Darwin in some passages seems to contemplate where, for example, he says that the female appreciates the display of the male and places to her credit a taste for the beautiful. But Darwin himself distinctly states[177] that "it is not probable that she consciously deliberates; but she is most excited or attracted by the most beautiful, or melodious, or gallant males." The view here put forward, which has been developed by Prof. Groos,[178] therefore seems to have Darwin's own sanction. The phenomena are not only biological; there are psychological elements as well. One can hardly suppose that the female is unconscious of the male's presence; the final yielding must surely be accompanied by heightened emotional tone. Whether we call it choice or not is merely a matter of definition of terms. The behaviour is in part determined by supplementary psychological values. Prof. Groos regards the coyness of females as "a most efficient means of preventing the too early and too frequent yielding to the sexual impulse."[179] Be that as it may, it is, in any case, if we grant the facts, a means through which male sexual behaviour with all its biological and psychological implications, is raised to a level otherwise perhaps unattainable by natural means, while in the female it affords opportunities for the development in the individual and evolution in the race of what we may follow Darwin in calling appreciation, if we empty this word of the aesthetic implications which have gathered round it in the mental life of man.

Regarded from this standpoint of sexual selection, broadly considered, has probably been of great importance. The psychological accompaniments of the pairing situation have profoundly influenced the course of biological evolution and are themselves the outcome of that evolution.

Darwin makes only passing reference to those modes of behaviour in animals which go by the name of play. "Nothing," he says,[180] "is more common than for animals to take pleasure in practising whatever instinct they follow at other times for some real good." This is one of the very numerous cases in which a hint of the master has served to stimulate research in his disciples. It was left to Prof. Groos to develop this subject on evolutionary lines and to elaborate in a masterly manner Darwin's suggestion. "The utility of play," he says,[181] "is incalculable. This utility consists in the practice and exercise it affords for some of the more important duties of life,"--that is to say, for the performance of activities which will in adult life be essential to survival. He urges[182] that "the play of young animals has its origin in the fact that certain very important instincts appear at a time when the animal does not seriously need them." It is, however, questionable whether any instincts appear at a time when they are not needed. And it is questionable whether the instinctive and emotional attitude of the play-fight, to take one example, can be identified with those which accompany fighting in earnest, though no doubt they are closely related and have some common factors. It is probable that play, as preparatory behaviour, differs in biological detail (as it almost certainly does in emotional attributes) from the earnest of after-life and that it has been evolved through differentiation and integration of the primary tissue of experience, as a preparation through which certain essential modes of skill may be acquired--those animals in which the preparatory play-propensity was not inherited in due force and requisite amount being subsequently eliminated in the struggle for existence. In any case there is little question that Prof. Groos is right in basing the play-propensity on instinctive foundations.[183] None the less, as he contends, the essential biological value of play is that it is a means of training the educable nerve-tissue, of developing that part of the brain which is modified by experience and which thus acquires new characters, of elaborating the secondary tissue of experience on the predetermined lines of instinctive differentiation and thus furthering the psychological activities which are included under the comprehensive term "intelligent."

In The Descent of Man Darwin dealt at some length with intelligence and the higher mental faculties.[184] His object, he says, is to show that there is no fundamental difference between man and the higher mammals in their mental faculties; that these faculties are variable and the variations tend to be inherited; and that under natural selection beneficial variations of all kinds will have been preserved and injurious ones eliminated.

Darwin was too good an observer and too honest a man to minimise the "enormous difference" between the level of mental attainment of civilised man and that reached by any animal. His contention was that the difference, great as it is, is one of degree and not of kind. He realised that, in the development of the mental faculties of man, new factors in evolution have supervened--factors which play but a subordinate and subsidiary part in animal intelligence. Intercommunication by means of language, approbation and blame, and all that arises out of reflective thought, are but foreshadowed in the mental life of animals. Still he contends that these may be explained on the doctrine of evolution. He urges[185] "that man is variable in body and mind; and that the variations are induced, either directly or indirectly, by the same general causes, and obey the same general laws, as with the lower animals." He correlates mental development with the evolution of the brain.[186] "As the various mental faculties gradually developed themselves, the brain would almost certainly become larger. No one, I presume, doubts that the large proportion which the size of man's brain bears to his body, compared to the same proportion in the gorilla or orang, is closely connected with his higher mental powers." "With respect to the lower animals," he says,[187] "M. E. Lartet,[188] by comparing the crania of tertiary and recent mammals belonging to the same groups, has come to the remarkable conclusion that the brain is generally larger and the convolutions are more complex in the more recent form."

Sir E. Ray Lankester has sought to express in the simplest terms the implications of the increase in size of the cerebrum. "In what," he asks, "does the advantage of a larger cerebral mass consist?" "Man," he replies, "is born with fewer ready-made tricks of the nerve-centres--these performances of an inherited nervous mechanism so often called by the ill-defined term 'instincts'--than are the monkeys or any other animal. Correlated with the absence of inherited ready-made mechanism, man has a greater capacity of developing in the course of his individual growth similar nervous mechanisms (similar to but not identical with those of 'instinct') than any other animal.... The power of being educated--'educability' as we may term it--is what man possesses in excess as compared with the apes. I think we are justified in forming the hypothesis that it is this 'educability' which is the correlative of the increased size of the cerebrum." There has been natural selection of the more educable animals, for "the character which we describe as 'educability' can be transmitted, it is a congenital character. But the results of education can not be transmitted. In each generation they have to be acquired afresh, and with increased 'educability' they are more readily acquired and a larger variety of them.... The fact is that there is no community between the mechanisms of instinct and the mechanisms of intelligence, and that the latter are later in the history of the evolution of the brain than the former and can only develop in proportion as the former become feeble and defective."[189]

In this statement we have a good example of the further development of views which Darwin foreshadowed but did not thoroughly work out. It states the biological case clearly and tersely. Plasticity of behaviour in special accommodation to special circumstances is of survival value; it depends upon acquired characters; it is correlated with increase in size and complexity of the cerebrum; under natural selection therefore the larger and more complex cerebrum as the organ of plastic behaviour has been the outcome of natural selection. We have thus the biological foundations for a further development of genetic psychology.

There are diversities of opinion, as Darwin showed, with regard to the range of instinct in man and the higher animals as contrasted with lower types. Darwin himself said[190] that "Man, perhaps, has somewhat fewer instincts than those possessed by the animals which come next to him in the series." On the other hand, Prof. Wm. James says[191] that man is probably the animal with most instincts. The true position is that man and the higher animals have fewer complete and self-sufficing instincts than those which stand lower in the scale of mental evolution, but that they have an equally large or perhaps larger mass of instinctive raw material which may furnish the stuff to be elaborated by intelligent processes. There is, perhaps, a greater abundance of the primary tissue of experience to be refashioned and integrated by secondary modification; there is probably the same differentiation in relation to the determining biological ends, but there is at the outset less differentiation of the particular and specific modes of behaviour. The specialised instinctive performances and their concomitant experience-complexes are at the outset more indefinite. Only through acquired connections, correlated with experience, do they become definitely organised.

The full working-out of the delicate and subtle relationship of instinct and educability--that is, of the hereditary and the acquired factors in the mental life--is the task which lies before genetic and comparative psychology. They interact throughout the whole of life, and their interactions are very complex. No one can read the chapters of The Descent of Man which Darwin devotes to a consideration of the mental characters of man and animals without noticing, on the one hand, how sedulous he is in his search for hereditary foundations, and, on the other hand, how fully he realises the importance of acquired habits of mind. The fact that educability itself has innate tendencies--is in fact a partially differentiated educability--renders the unravelling of the factors of mental progress all the more difficult.

In his comparison of the mental powers of men and animals it was essential that Darwin should lay stress on points of similarity rather than on points of difference. Seeking to establish a doctrine of evolution, with its basal concept of continuity of process and community of character, he was bound to render clear and to emphasise the contention that the difference in mind between man and the higher animals, great as it is, is one of degree and not of kind. To this end Darwin not only recorded a large number of valuable observations of his own, and collected a considerable body of information from reliable sources, he presented the whole subject in a new light and showed that a natural history of mind might be written and that this method of study offered a wide and rich field for investigation. Of course those who regarded the study of mind only as a branch of metaphysics smiled at the philosophical ineptitude of the mere man of science. But the investigation, on natural history lines, has been prosecuted with a large measure of success. Much indeed still remains to be done; for special training is required, and the workers are still few. Promise for the future is however afforded by the fact that investigation is prosecuted on experimental lines and that something like organised methods of research are taking form. There is now but little reliance on casual observations recorded by those who have not undergone the necessary discipline in these methods. There is also some change of emphasis in formulating conclusions. Now that the general evolutionary thesis is fully and freely accepted by those who carry on such researches, more stress is laid on the differentiation of the stages of evolutionary advance than on the fact of their underlying community of nature. The conceptual intelligence which is especially characteristic of the higher mental procedure of man is more firmly distinguished from the perceptual intelligence which he shares with the lower animals--distinguished now as a higher product of evolution, no longer as differing in origin or different in kind. Some progress has been made, on the one hand in rendering an account of intelligent profiting by experience under the guidance of pleasure and pain in the perceptual field, on lines predetermined by instinctive differentiation for biological ends, and on the other hand in elucidating the method of conceptual thought employed, for example, by the investigator himself in interpreting the perceptual experience of the lower animals.

Thus there is a growing tendency to realise more fully that there are two orders of educability--first an educability of the perceptual intelligence based on the biological foundation of instinct, and secondly an educability of the conceptual intelligence which refashions and rearranges the data afforded by previous inheritance and acquisition. It is in relation to this second and higher order of educability that the cerebrum of man shows so large an increase of mass and a yet larger increase of effective surface through its rich convolutions. It is through educability of this order that the human child is brought intellectually and affectively into touch with the ideal constructions by means of which man has endeavoured, with more or less success, to reach an interpretation of nature, and to guide the course of the further evolution of his race--ideal constructions which form part of man's environment.

It formed no part of Darwin's purpose to consider, save in broad outline, the methods, or to discuss in any fulness of detail the results of the process by which a differentiation of the mental faculties of man from those of the lower animals has been brought about--a differentiation the existence of which he again and again acknowledges. His purpose was rather to show that, notwithstanding this differentiation, there is basal community in kind. This must be remembered in considering his treatment of the biological foundations on which man's systems of ethics are built. He definitely stated that he approached the subject "exclusively from the side of natural history."[192] His general conclusion is that the moral sense is fundamentally identical with the social instincts, which have been developed for the good of the community; and he suggests that the concept which thus enables us to interpret the biological ground-plan of morals also enables us to frame a rational ideal of the moral end. "As the social instincts," he says,[193] "both of man and the lower animals have no doubt been developed by nearly the same steps, it would be advisable, if found practicable, to use the same definition in both cases, and to take as the standard of morality, the general good or welfare of the community, rather than the general happiness." But the kind of community for the good of which the social instincts of animals and primitive men were biologically developed may be different from that which is the product of civilisation, as Darwin no doubt realised. Darwin's contention was that conscience is a social instinct and has been evolved because it is useful to the tribe in the struggle for existence against other tribes. One the other hand J. S. Mill urged that the moral feelings are not innate but acquired, and Bain held the same view, believing that the moral sense is acquired by each individual during his life-time. Darwin, who notes[194] their opinion with his usual candour, adds that "on the general theory of evolution this is at least extremely improbable." It is impossible to enter into the question here: much turns on the exact connotation of the terms "conscience" and "moral sense," and on the meaning we attach to the statement that the moral sense is fundamentally identical with the social instincts.

Presumably the majority of those who approach the subjects discussed in the third, fourth, and fifth chapters of The Descent of Man in the full conviction that mental phenomena, not less than organic phenomena, have a natural genesis, would, without hesitation, admit that the intellectual and moral systems of civilised man are ideal constructions, the products of conceptual thought, and that as such they are, in their developed form, acquired. The moral sentiments are the emotional analogues of highly developed concepts. This does not however imply that they are outside the range of natural history treatment. Even though it may be desirable to differentiate the moral conduct of men from the social behaviour of animals (to which some such term as "pre-moral" or "quasi-moral" may be applied), still the fact remains that, as Darwin showed, there is abundant evidence of the occurrence of such social behaviour--social behaviour which, even granted that it is in large part intelligently acquired, and is itself so far a product of educability, is of survival value. It makes for that integration without which no social group could hold together and escape elimination. Furthermore, even if we grant that such behaviour is intelligently acquired, that is to say arises through the modification of hereditary instincts and emotions, the fact remains that only through these instinctive and emotional data is afforded the primary tissue of the experience which is susceptible of such modification.

Darwin sought to show, and succeeded in showing, that for the intellectual and moral life there are instinctive foundations which a biological treatment alone can disclose. It is true that he did not in all cases analytically distinguish the foundations from the superstructure. Even to-day we are scarcely in a position to do so adequately. But his treatment was of great value in giving an impetus to further research. This value indeed can scarcely be over-estimated. And when the natural history of the mental operations shall have been written, the cardinal fact will stand forth, that the instinctive and emotional foundations are the outcome of biological evolution and have been ingrained in the race through natural selection. We shall more clearly realise that educability itself is a product of natural selection, though the specific results acquired through cerebral modifications are not transmitted through heredity.

It will, perhaps, also be realised that the instinctive foundations of social behaviour are, for us, somewhat out of date and have undergone but little change throughout the progress of civilisation, because natural selection has long since ceased to be the dominant factor in human progress. The history of human progress has been mainly the history of man's higher educability, the products of which he has projected on to his environment. This educability remains on the average what it was a dozen generations ago; but the thought-woven tapestry of his surroundings is refashioned and improved by each succeeding generation. Few men have in greater measure enriched the thought-environment with which it is the aim of education to bring educable human beings into vital contact, than has Charles Darwin. His special field of work was the wide province of biology; but he did much to help us to realise that mental factors have contributed to organic evolution and that in man, the highest product of Evolution, they have reached a position of unquestioned supremacy.

FOOTNOTES:

[Footnote 153: Origin of Species (6th edit.), p. 205.]

[Footnote 154: Descent of man (2nd edit. 1888), Vol. I. p. 99; Popular edit. p. 99.]

[Footnote 155: Ibid. p. 99.]

[Footnote 156: The Expression of the Emotions (2nd edit.), p. 32.]

[Footnote 157: Descent of Man, Vol. II. p. 435.]

[Footnote 158: Ibid. 437, 438.]

[Footnote 159: Origin of Species (6th edit.), p. 205.]

[Footnote 160: Origin of Species (6th edit.), p. 205.]

[Footnote 161: Ibid. p. 233.]

[Footnote 162: Ibid. p. 205.]

[Footnote 163: Origin of Species (6th edit.), p. 233.]

[Footnote 164: Origin of Species, pp. 210, 211.]

[Footnote 165: Independently suggested, on somewhat different lines, by Profs. J. Mark Baldwin, Henry F. Osborn and the writer.]

[Footnote 166: Origin of Species (6th edit.), p. 206.]

[Footnote 167: Expression of the Emotions, p. 13. The passage is here somewhat condensed.]

[Footnote 168: Ibid. p. 368.]

[Footnote 169: Expression of the Emotions, pp. 373, 374.]

[Footnote 170: Ibid. p. 368.]

[Footnote 171: Ibid. p. 369.]

[Footnote 172: Expression of the Emotions, pp. 65 ff.]

[Footnote 173: Cf. William James, Principles of Psychology, Vol. II. Chap. XXV, New York, 1890.]

[Footnote 174: Darwinism, pp. 282, 283, London, 1889.]

[Footnote 175: Ibid. p. 283.]

[Footnote 176: Darwinism, pp. 283, 284.]

[Footnote 177: Descent of Man (2nd edit.), Vol. II. pp. 136, 137; (Popular edit.), pp. 642, 643.]

[Footnote 178: The Play of Animals, p. 244, London, 1898.]

[Footnote 179: Ibid. p. 283.]

[Footnote 180: Descent of Man, Vol. II. p. 60; (Popular edit.), p. 566.]

[Footnote 181: The Play of Animals, p. 76.]

[Footnote 182: Ibid. p. 75.]

[Footnote 183: The Play of Animals p. 24.]

[Footnote 184: Descent of Man (1st edit.), Chaps. II, III, V; (2nd edit.), Chaps. III, IV, V.]

[Footnote 185: Descent of Man, Vol. I. pp. 70, 71; (Popular edit.), pp. 70, 71.]

[Footnote 186: Ibid. p. 81.]

[Footnote 187: Ibid. (Popular edit.), p. 82.]

[Footnote 188: Comptes Rendus des Sciences, June 1, 1868.]

[Footnote 189: Nature, Vol. LXI. pp. 624, 625 (1900).]

[Footnote 190: Descent of Man, Vol. I. p. 100.]

[Footnote 191: Principles of Psychology, Vol. II. p. 289.]

[Footnote 192: Descent of Man, Vol. I. p. 149.]

[Footnote 193: Descent of Man, p. 185.]

[Footnote 194: Ibid. p. 150 (footnote).]

VII

THE INFLUENCE OF THE CONCEPTION OF EVOLUTION ON MODERN PHILOSOPHY

BY H. HAFDING

Professor of Philosophy in the University of Copenhagen

I

It is difficult to draw a sharp line between philosophy and natural science. The naturalist who introduces a new principle, or demonstrates a fact which throws a new light on existence, not only renders an important service to philosophy but is himself a philosopher in the broader sense of the word. The aim of philosophy in the stricter sense is to attain points of view from which the fundamental phenomena and the principles of the special sciences can be seen in their relative importance and connection. But philosophy in this stricter sense has always been influenced by philosophy in the broader sense. Greek philosophy came under the influence of logic and mathematics, modern philosophy under the influence of natural science. The name of Charles Darwin stands with those of Galileo, Newton, and Robert Mayer-- names which denote new problems and great alterations in our conception of the universe.

First of all we must lay stress on Darwin's own personality. His deep love of truth, his indefatigable inquiry, his wide horizon, and his steady self-criticism make him a scientific model, even if his results and theories should eventually come to possess mainly an historical interest. In the intellectual domain the primary object is to reach high summits from which wide surveys are possible, to reach them toiling honestly upwards by the way of experience, and then not to turn dizzy when a summit is gained. Darwinians have sometimes turned dizzy, but Darwin never. He saw from the first the great importance of his hypothesis, not only because of its solution of the old problem as to the value of the concept of species, not only because of the grand picture of natural evolution which it unrolls, but also because of the life and inspiration its method would impart to the study of comparative anatomy, of instinct and of heredity, and finally because of the influence it would exert on the whole conception of existence. He wrote in his note-book in the year 1837: "My theory would give zest to recent and fossil comparative anatomy; it would lead to the study of instinct, heredity, and mind-heredity, whole [of] metaphysics."[195]

We can distinguish four main points in which Darwin's investigations possess philosophical importance.

The evolution hypothesis is much older than Darwin; it is, indeed, one of the oldest guessings of human thought. In the eighteenth century is was put forward by Diderot and Lamettrie and suggested by Kant (1786). As we shall see later, it was held also by several philosophers in the first half of the nineteenth century. In his preface to The Origin of Species, Darwin mentions the naturalists who were his forerunners. But he has set forth the hypothesis of evolution in so energetic and thorough a manner that it perforce attracts the attention of all thoughtful men in a much higher degree than it did before the publication of the Origin.

And further, the importance of his teaching rests on the fact that he, much more than his predecessors, even than Lamarck, sought a foundation for his hypothesis in definite facts. Modern science began by demanding--with Kepler and Newton--evidence of varae causae; this demand Darwin industriously set himself to satisfy--hence the wealth of material which he collected by his observations and his experiments. He not only revived an old hypothesis, but he saw the necessity of verifying it by facts. Whether the special cause on which he founded the explanation of the origin of species-- Natural Selection--is sufficient, is now a subject of discussion. He himself had some doubt in regard to this question, and the criticisms which are directed against his hypothesis hit Darwinism rather than Darwin. In his indefatigable search for empirical evidence he is a model even for his antagonists: he has compelled them to approach the problems of life along other lines than those which were formerly followed.

Whether the special cause to which Darwin appealed is sufficient or not, at least to it is probably due the greater part of the influence which he has exerted on the general trend of thought. "Struggle for existence" and "natural selection" are principles which have been applied, more or less, in every department of thought. Recent research, it is true, has discovered greater empirical discontinuity--leaps, "mutations"--whereas Darwin believed in the importance of small variations slowly accumulated. It has also been shown by the experimental method, which in recent biological work has succeeded Darwin's more historical method, that types once constituted possess great permanence, the fluctuations being restricted within clearly defined boundaries. The problem has become more precise, both as to variation and as to heredity. The inner conditions of life have in both respects shown a greater independence than Darwin had supposed in his theory, though he always admitted that the cause of variation was to him a great enigma, "a most perplexing problem," and that the struggle for life could only occur where variation existed. But, at any rate, it was of the greatest importance that Darwin gave a living impression of the struggle for life which is everywhere going on, and to which even the highest forms of existence must be amenable. The philosophical importance of these ideas does not stand or fall with the answer to the question, whether natural selection is a sufficient explanation of the origin of species or not it has an independent, positive value for everyone who will observe life and reality with an unbiased mind.

In accentuating the struggle for life Darwin stands as a characteristically English thinker: he continues a train of ideas which Hobbes and Malthus had already begun. Moreover in his critical views as to the conception of species he had English forerunners; in the middle ages Occam and Duns Scotus, in the eighteenth century Berkeley and Hume. In his moral philosophy, as we shall see later, he is an adherent of the school which is represented by Hutcheson, Home and Adam Smith. Because he is no philosopher in the stricter sense of the term, it is of great interest to see that his attitude of mind is that of the great thinkers of his nation.

In considering Darwin's influence on philosophy we will begin with an examination of the attitude of philosophy to the conception of evolution at the time when The Origin of Species appeared. We will then examine the effects which the theory of evolution, and especially the idea of the struggle for life, has had, and naturally must have, on the discussion of philosophical problems.

II

When The Origin of Species appeared fifty years ago Romantic speculation, Schelling's and Hegel's philosophy, still reigned on the continent, while in England Positivism, the philosophy of Comte and Stuart Mill, represented the most important trend of thought. German speculation had much to say on evolution, it even pretended to be a philosophy of evolution. But then the word "evolution" was to be taken in an ideal, not in a real, sense. To speculative thought the forms and types of nature formed a system of ideas, within which any form could lead us by continuous transitions to any other. It was a classificatory system which was regarded as a divine world of thought or images, within which metamorphoses could go on--a condition comparable with that in the mind of the poet when one image follows another with imperceptible changes. Goethe's ideas of evolution, as expressed in his Metamorphosen der Pflanzen und der Thiere, belong to this category; it is, therefore, incorrect to call him a forerunner of Darwin. Schelling and Hegel held the same idea; Hegel expressly rejected the conception of a real evolution in time as coarse and materialistic. "Nature," he says, "is to be considered as a system of stages, the one necessarily arising from the other, and being the nearest truth of that from which it proceeds; but not in such a way that the one is naturally generated by the other; on the contrary [their connection lies] in the inner idea which is the ground of nature. The metamorphosis can be ascribed only to the notion as such, because it alone is evolution.... It has been a clumsy idea in the older as well as in the newer philosophy of nature, to regard the transformation and the transition from one natural form and sphere to a higher as an outward and actual production."[196]

The only one of the philosophers of Romanticism who believed in a real, historical evolution, a real production of new species, was Oken.[197] Danish philosophers, such as Treschow (1812) and Sibbern (1846), have also broached the idea of an historical evolution of all living beings from the lowest to the highest. Schopenhauer's philosophy has a more realistic character than that of Schelling's and Hegel's, his diametrical opposites, although he also belongs to the romantic school of thought. His philosophical and psychological views were greatly influenced by French naturalists and philosophers, especially by Cabanis and Lamarck. He praises the "ever memorable Lamarck," because he laid so much stress on the "will to live." But he repudiates as a "wonderful error" the idea that the organs of animals should have reached their present perfection through a development in time, during the course of innumerable generations. It was, he said, a consequence of the low standard of contemporary French philosophy, that Lamarck came to the idea of the construction of living beings in time through succession![198]

The positivistic stream of thought was not more in favour of a real evolution than was the Romantic school. Its aim was to adhere to positive facts: it looked with suspicion on far-reaching speculation. Comte laid great stress on the discontinuity found between the different kingdoms of nature, as well as within each single kingdom. As he regarded as unscientific every attempt to reduce the number of physical forces, so he rejected entirely the hypothesis of Lamarck concerning the evolution of species; the idea of species would in his eyes absolutely lose its importance if a transition from species to species under the influence of conditions of life were admitted. His disciples (Littr? Robin) continued to direct against Darwin the polemics which their master had employed against Lamarck. Stuart Mill, who, in the theory of knowledge, represented the empirical or positivistic movement in philosophy--like his English forerunners from Locke to Hume--founded his theory of knowledge and morals on the experience of the single individual. He sympathised with the theory of the original likeness of all individuals and derived their differences, on which he practically and theoretically laid much stress, from the influence both of experience and education, and, generally, of physical and social causes. He admitted an individual evolution, and, in the human species, an evolution based on social progress; but no physiological evolution of species. He was afraid that the hypothesis of heredity would carry us back to the old theory of "innate" ideas.

Darwin was more empirical than Comte and Mill; experience disclosed to him a deeper continuity than they could find; closer than before the nature and fate of the single individual were shown to be interwoven in the great web binding the life of the species with nature as a whole. And the continuity which so many idealistic philosophers could find only in the world of thought, he showed to be present in the world of reality.

III

Darwin's energetic renewal of the old idea of evolution has its chief importance in strengthening the conviction of this real continuity in the world, of continuity in the series of form and events. It was a great support for all those who were prepared to base their conception of life on scientific grounds. Together with the recently discovered law of the conservation of energy, it helped to produce the great realistic movement which characterises the last third of the nineteenth century. After the decline of the Romantic movement people wished to have firmer ground under their feet and reality now asserted itself in a more emphatic manner than in the period of Romanticism. It was easy for Hegel to proclaim that "the real" was "the rational," and that "the rational" was "the real": reality itself existed for him only in the interpretation of ideal reason, and if there was anything which could not be merged in the higher unity of thought, then it was only an example of the "impotence of nature to hold to the idea." But now concepts are to be founded on nature and not on any system of categories too confidently deduced ?priori. The new devotion to nature had its recompense in itself, because the new points of view made us see that nature could indeed "hold to ideas," though perhaps not to those which we had cogitated beforehand.

A most important question for philosophers to answer was whether the new views were compatible with an idealistic conception of life and existence. Some proclaimed that we have now no need of any philosophy beyond the principles of the conservation of matter and energy and the principle of natural evolution: existence should and could be definitely and completely explained by the laws of material nature. But abler thinkers saw that the thing was not so simple. They were prepared to give the new views their just place and to examine what alterations the old views must undergo in order to be brought into harmony with the new data.

The realistic character of Darwin's theory was shown not only in the idea of natural continuity, but also, and not least, in the idea of the cause whereby organic life advances step by step. This idea--the idea of the struggle for life-- implied that nothing could persist, if it had no power to maintain itself under the given conditions. Inner value alone does not decide. Idealism was here put to its hardest trial. In continuous evolution it could perhaps still find an analogy to the inner evolution of ideas in the mind; but in the demand for power in order to struggle with outward conditions Realism seemed to announce itself in its most brutal form. Every form of Idealism had to ask itself seriously how it was going to "struggle for life" with this new Realism.

We will now give a short account of the position which leading thinkers in different countries have taken up in regard to this question.

I. Herbert Spencer was the philosopher whose mind was best prepared by his own previous thinking to admit the theory of Darwin to a place in his conception of the world. His criticism of the arguments which had been put forward against the hypothesis of Lamarck, showed that Spencer, as a young man, was an adherent to the evolution idea. In his Social Statics (1850) he applied this idea to human life and moral civilisation. In 1852 he wrote an essay on The Development Hypothesis, in which he definitely stated his belief that the differentiation of species, like the differentiation within a single organism, was the result of development. In the first edition of his Psychology (1855) he took a step which put him in opposition to the older English school (from Locke to Mill): he acknowledged "innate ideas" so far as to admit the tendency of acquired habits to be inherited in the course of generations, so that the nature and functions of the individual are only to be understood through its connection with the life of the species. In 1857, in his essay on Progress, he propounded the law of differentiation as a general law of evolution, verified by examples from all regions of experience, the evolution of species being only one of these examples. On the effect which the appearance of The Origin of Species had on his mind he writes in his Autobiography: "Up to that time ... I held that the sole cause of organic evolution is the inheritance of functionally-produced modifications. The Origin of Species made it clear to me that I was wrong, and that the larger part of the facts cannot be due to any such cause.... To have the theory of organic evolution justified was of course to get further support for that theory of evolution at large with which ... all my conceptions were bound up."[199] Instead of the metaphorical expression "natural selection," Spencer introduced the term "survival of the fittest," which found favour with Darwin as well as with Wallace.

In working out his ideas of evolution, Spencer found that differentiation was not the only form of evolution. In its simplest form evolution is mainly a concentration, previously scattered elements being integrated and losing independent movement. Differentiation is only forthcoming when minor wholes arise within a greater whole. And the highest form of evolution is reached when there is a harmony between concentration and differentiation, a harmony which Spencer calls equilibration and which he defines as a moving equilibrium. At the same time this definition enables him to illustrate the expression "survival of the fittest." "Every living organism exhibits such a moving equilibrium--a balanced set of functions constituting its life; and the overthrow of this balanced set of functions or moving equilibrium is what we call death. Some individuals in a species are so constituted that their moving equilibria are less easily overthrown than those of other individuals; and these are the fittest which survive, or, in Mr. Darwin's language, they are the select which nature preserves."[200] Not only in the domain of organic life, but in all domains, the summit of evolution is, according to Spencer, characterised by such a harmony--by a moving equilibrium.

Spencer's analysis of the concept of evolution, based on a great variety of examples, has made this concept clearer and more definite than before. It contains the three elements; integration, differentiation and equilibration. It is true that a concept which is to be valid for all domains of experience must have an abstract character, and between the several domains there is, strictly speaking, only a relation of analogy. So there is only analogy between psychical and physical evolution. But this is no serious objection, because general concepts do not express more than analogies between the phenomena which they represent. Spencer takes his leading forms from the material world in defining evolution (in the simplest form) as integration of matter and dissipation of movement; but as he--not always quite consistently[201]--assumed a correspondence of mind and matter, he could very well give these terms an indirect importance for psychical evolution. Spencer has always, in my opinion with full right, repudiated the ascription of materialism. He is no more a materialist than Spinoza. In his Principles of Psychology (?63) he expressed himself very clearly: "Though it seems easier to translate so-called matter into so-called spirit, than to translate so-called spirit into so-called matter--which latter is indeed wholly impossible--yet no translation can carry us beyond our symbols." These words lead us naturally to a group of thinkers whose starting-point was psychical evolution. But we have still one aspect of Spencer's philosophy to mention.

Spencer founded his "laws of evolution" on an inductive basis, but he was convinced that they could be deduced from the law of the conservation of energy. Such a deduction is, perhaps, possible for the more elementary forms of evolution, integration and differentiation; but it is not possible for the highest form, the equilibration, which is a harmony of integration and differentiation. Spencer can no more deduce the necessity for the eventual appearance of "moving equilibria" of harmonious totalities than Hegel could guarantee the "higher unities" in which all contradictions should be reconciled. In Spencer's hands the theory of evolution acquired a more decidedly optimistic character than in Darwin's; but I shall deal later with the relation of Darwin's hypothesis to the opposition of optimism and pessimism.

II. While the starting-point of Spencer was biological or cosmological, psychical evolution being conceived as in analogy with physical, a group of eminent thinkers--in Germany Wundt, in France Fouille, in Italy Ardig-took, each in his own manner, their starting-point in psychical evolution as an original fact and as a type of all evolution, the hypothesis of Darwin coming in as a corroboration and as a special example. They maintain the continuity of evolution; they find this character most prominent in psychical evolution, and this is for them a motive to demand a corresponding continuity in the material, especially in the organic domain.

To Wundt the concept of will is prominent. They see the type of all evolution in the transformation of the life of will from blind impulse to conscious choice; the theories of Lamarck and Darwin are used to support the view that there is in nature a tendency to evolution in steady reciprocity with external conditions. The struggle for life is here only a secondary fact. Its apparent prominence is explained by the circumstance that the influence of external conditions is easily made out, while inner conditions can be verified only through their effects. For Ardig?the evolution of thought was the starting-point and the type: in the evolution of a scientific hypothesis we see a progress from the indefinite (indistinto) to the definite (distinto), and this is a characteristic of all evolution, as Ardig?has pointed out in a series of works. The opposition between indistinto and distinto corresponds to Spencer's opposition between homogeneity and heterogeneity. The hypothesis of the origin of differences of species from more simple forms is a special example of the general law of evolution.

In the views of Wundt we find the fundamental idea of idealism psychical phenomena as expressions of the innermost nature of existence. They differ from the older Idealism in the great stress which they lay on evolution as a real, historical process which is going on through steady conflict with external conditions. The Romantic dread of reality is broken. It is beyond doubt that Darwin's emphasis on the struggle for life as a necessary condition of evolution has been a very important factor in carrying philosophy back to reality from the heaven of pure ideas. The philosophy of Ardig? on the other side, appears more as a continuation and deepening of positivism, though the Italian thinker arrived at his point of view independently of French-English positivism. The idea of continuous evolution is here maintained in opposition to Comte's and Mill's philosophy of discontinuity. From Wundt in conceiving psychical evolution not as an immediate revelation of the innermost nature of existence, but only as a single, though the most accessible example, of evolution.

III. To the French philosophers Boutroux and Bergson, evolution proper is continuous and qualitative, while outer experience and physical science give us fragments only, sporadic processes and mechanical combinations. To Bergson, in his recent work L'Evolution Creatrice, evolution consists in an de vie which to our fragmentary observation and analytic reflexion appears as broken into a manifold of elements and processes. The concept of matter in its scientific form is the result of this breaking asunder, essential for all scientific reflexion. In these conceptions the strongest opposition between inner and outer conditions of evolution is expressed: in the domain of internal conditions spontaneous development of qualitative forms--in the domain of external conditions discontinuity and mechanical combination.

We see, then, that the theory of evolution has influenced philosophy in a variety of forms. It has made idealistic thinkers revise their relation to the real world; it has led positivistic thinkers to find a closer connection between the facts on which they based their views; it has made us all open our eyes for new possibilities to arise through the prima facie inexplicable "spontaneous" variations which are the condition of all evolution. This last point is one of peculiar interest. Deeper than speculative philosophy and mechanical science saw in the days of their triumph, we catch sight of new streams, whose sources and laws we have still to discover. Most sharply does this appear in the theory of mutation, which is only a stronger accentuation of a main point in Darwinism. It is interesting to see that an analogous problem comes into the foreground in physics through the discovery of radioactive phenomena, and in psychology through the assumption of psychical new formations (as held by Boutroux, William James and Bergson). From this side, Darwin's ideas, as well as the analogous ideas in other domains, incite us to renewed examination of our first principles, their rationality and their value. On the other hand, his theory of the struggle for existence challenges us to examine the conditions and discuss the outlook as to the persistence of human life and society and of the values that belong to them. It is not enough to hope (or fear?) the rising of new forms; we have also to investigate the possibility of upholding the forms and ideals which have hitherto been the bases of human life. Darwin has here given his age the most earnest and most impressive lesson. This side of Darwin's theory is of peculiar interest to some special philosophical problems to which I now pass.

IV

Among philosophical problems the problem of knowledge has in the last century occupied a foremost place. It is natural, then, to ask how Darwin and the hypothesis whose most eminent representative he is, stand to this problem.

Darwin started an hypothesis. But every hypothesis is won by inference from certain presuppositions, and every inference is based on the general principles of human thought. The evolution hypothesis presupposes, then, human thought and its principles. And not only the abstract logical principles are thus pre-supposed. The evolution hypothesis purports to be not only a formal arrangement of phenomena, but to express also the law of a real process. It supposes, then, that the real data--all that in our knowledge which we do not produce ourselves, but which we in the main simply receive--are subject to laws which are at least analogous to the logical relations of our thoughts; in other words, it assumes the validity of the principle of causality. If organic species could arise without cause there would be no use in framing hypotheses. Only if we assume the principle of causality, is there a problem to solve.

Though Darwinism has had a great influence on philosophy considered as a striving after a scientific view of the world, yet here is a point of view--the epistemological--where philosophy is not only independent but reaches beyond any result of natural science. Perhaps it will be said: the powers and functions of organic beings only persist (perhaps also only arise) when they correspond sufficiently to the conditions under which the struggle of life is to go on. Human thought itself is, then, a variation (or a mutation) which has been able to persist and to survive. Is not, then, the problem of knowledge solved by the evolution hypothesis? Spencer had given an affirmative answer to this question before the appearance of The Origin of Species. For the individual, he said, there is an priori, original, basis (or Anlage) for all mental life; but in the species all powers have developed in reciprocity with extendal conditions. Knowledge is here considered from the practical point of view, as a weapon in the struggle for life, as an "organon" which has been continuously in use for generations. In recent years the economic or pragmatic epistemology, as developed by Avenarius and Mach in Germany, and by James in America, points in the same direction. Science, it is said, only maintains those principles and presuppositions which are necessary to the simplest and clearest orientation be applied to experience and to practical work, will successively be eliminated.

In these views a striking and important application is made of the idea of struggle for life to the development of human thought. Thought must, as all other things in the world, struggle for life. But this whole consideration belongs to psychology, not to the theory of knowledge (epistemology), which is concerned only with the validity of knowledge, not with its historical origin. Every hypothesis to explain the origin of knowledge must submit to cross-examination by the theory of knowledge, because it works with the fundamental forms and principles of human thought. We cannot go further back than these forms and principles, which it is the aim of epistemology to ascertain and for which no further reason can be given.[202]

But there is another side of the problem which is, perhaps, of more importance and which epistemology generally overlooks. If new variations can arise, not only in organic but perhaps also in inorganic nature, new tasks are placed before the human mind. The question is, then, if it has forms in which there is room for the new matter? We are here touching a possibility which the great master of epistemology did not bring to light. Kant supposed confidently that no other matter of knowledge could stream forth from the dark source which he called "the thing-in-itself," than such as could be synthesised in our existing forms of knowledge. He mentions the possibility of other forms than the human, and warns us against the dogmatic assumption that the human conception of existence should be absolutely adequate. But he seems to be quite sure that the thing-in-itself works constantly, and consequently always gives us only what our powers can master. This assumption was a consequence of Kant's rationalistic tendency, but one for which no warrant can be given. Evolutionism and systematism are opposing tendencies which can never be absolutely harmonised one with the other. Evolution may at any time break some form which the system-monger regards as finally established. Darwin himself felt a great difference in looking at variation as an evolutionist and as a systematist. When he was working at his evolution theory, he was very glad to find variations; but they were a hindrance to him when he worked as a systematist, in preparing his work on Cirripedia. He says in a letter: "I had thought the same parts of the same species more resemble (than they do anyhow in Cirripedia) objects cast in the same mould. Systematic work would be easy were it not for this confounded variation, which, however, is pleasant to me as a speculatist, though odious to me as a systematist."[203] He could indeed be angry with variations even as an evolutionist; but then only because he could not explain them, not because he could not classify them. "If, as I must think, external conditions produce little direct effect, what the devil determines each particular variation?"[204] What Darwin experienced in this particular domain holds good of all knowledge. All knowledge is systematic, in so far as it strives to put phenomena in quite definite relations, one to another. But the systematisation can never be complete. And here Darwin has contributed much to widen the world, for us. He has shown us forces and tendencies in nature which make absolute systems impossible, at the same time that they give us new objects and problems. There is still a place for what Lessing called "the unceasing striving after truth," while "absolute truth" (in the sense of a closed system) is unattainable so long as life and experience are going on.

There is here a special remark to be made. As we have seen above, recent research has shown that natural selection or struggle for life is no explanation of variations. Hugo de Vries distinguishes between partial and embryonal variations, or between variations and mutations, only the last-named being heritable, and therefore of importance for the origin of new species. But the existence of variations is not only of interest for the problem of the origin of species; it has also a more general interest. An individual does not lose its importance for knowledge, because its qualities are not heritable. On the contrary, in higher beings at least, individual peculiarities will become more and more independent objects of interest. Knowledge takes account of the biographies not only of species, but also of individuals: it seeks to find the law of development of the single individual.[205] As Leibnitz said long ago, individuality consists in the law of the changes of a being: "La loi du changement fait l'individualit?de chaque substance." Here is a world which is almost new for science, which till now has mainly occupied itself with general laws and forms. But these are ultimately only means to understand the individual phenomena, in whose nature and history a manifold of laws and forms always cooperate. The importance of this remark will appear in the sequel.

V

To many people the Darwinian theory of natural selection or struggle for existence seemed to change the whole conception of life, and particularly all the conditions on which the validity of ethical ideas depends. If only that has persistence which can be adapted to a given condition, what will then be the fate of our ideals, of our standards of good and evil? Blind force seems to reign, and the only thing that counts seems to be the most heedless use of power. Darwinism, it was said, has proclaimed brutality. No other difference seems permanent save that between the sound, powerful and happy on the one side, the sick, feeble and unhappy on the other; and every attempt to alleviate this difference seems to lead to general enervation. Some of those who interpreted Darwinism in this manner felt an aesthetic delight in contemplating the heedlessness and energy of the great struggle for existence and anticipated the realisation of a higher human type as the outcome of it: so Nietzsche and his followers. Others recognising the same consequences in Darwinism regarded these as one of the strongest objections against it; so Daring and Kropotkin (in his earlier works).

This interpretation of Darwinism was frequent in the interval between the two main works of Darwin--The Origin of Species and The Descent of Man. But even during this interval it was evident to an attentive reader that Darwin himself did not found his standard of good and evil on the features of the life of nature he had emphasised so strongly. He did not justify the ways along which nature reached its ends; he only pointed them out. The "real" was not to him, as to Hegel, one with the "rational." Darwin has, indeed, by his whole conception of nature, rendered a great service to ethics in making the difference between the life of nature and the ethical life appear in so strong a light. The ethical problem could now be stated in a sharper form than before. But this was not the first time that the idea of the struggle for life was put in relation to the ethical problem. In the seventeenth century Thomas Hobbes gave the first impulse to the whole modern discussion of ethical principles in his theory of bellum omnium contra omnes. Men, he taught, are in the state of nature enemies one of another, and they live either in fright or in the glory of power. But it was not the opinion of Hobbes that this made ethics impossible. On the contrary, he found a standard for virtue and vice in the fact that some qualities and actions have a tendency to bring us out of the state of war and to secure peace, while other qualities have a contrary tendency. In the eighteenth century even Immanuel Kant's ideal ethics had--so far as can be seen--a similar origin. Shortly before the foundation of his definitive ethics, Kant wrote his Idee zu einer allgemeinen Weltgeschichte (1784), where--in a way which reminds us of Hobbes, and is prophetic of Darwin--he describes the forward-driving power of struggle in the human world. It is here as with the struggle of the trees for light and air, through which they compete with one another in height. Anxiety about war can only be allayed by an ordinance which gives everyone his full liberty under acknowledgment of the equal liberty of others. And such ordinance and acknowledgment are also attributes of the content of the moral law, as Kant proclaimed it in the year after the publication of his essay (1785).[206] Kant really came to his ethics by the way of evolution, though he afterwards disavowed it. Similarly the same line of thought may be traced in Hegel though it has been disguised in the form of speculative dialectics.[207] And in Schopenhauer's theory of the blind will to live and its abrogation by the ethical feeling, which is founded on universal sympathy, we have a more individualistic form of the same idea.

It was, then, not entirely a foreign point of view which Darwin introduced into ethical thought, even if we take no account of the poetical character of the word "struggle" and of the more direct adaptation, through the use and non-use of power, which Darwin also emphasised. In The Descent of Man he has devoted a special chapter[208] to a discussion of the origin of the ethical consciousness. The characteristic expression of this consciousness he found, just as Kant did, in the idea of "ought"; it was the origin of this new idea which should be explained. His hypothesis was that the ethical "ought" has its origin in the social and parental instincts, which, as well as other instincts (e.g. the instinct of self-preservation), lie deeper than pleasure and pain. In many species, not least in the human species, these instincts are fostered by natural selection; and when the powers of memory and comparison are developed, so that single acts can be valued according to the claims of the deep social instinct, then consciousness of duty and remorse are possible. Blind instinct has developed to conscious ethical will.

As already stated, Darwin, as a moral philosopher belongs to the school that was founded by Shaftesbury, and was afterwards represented by Hutcheson,

Hume, Adam Smith, Comte and Spencer. His merit is, first, that he has given this tendency of thought a biological foundation, and that he has stamped on it a doughty character in showing that ethical ideas and sentiments, rightly conceived, are forces which are at work in the struggle for life.

There are still many questions to solve. Not only does the ethical development within the human species contain features still unexplained;[209] but we are confronted by the great problem whether after all a genetic historical theory can be of decisive importance here. To every consequent ethical consciousness there is a standard of value, a primordial value which determines the single ethical judgments as their last presupposition, and the "rightness" of this basis, the "value" of this value can as little be discussed as the "rationality" of our logical principles. There is here revealed a possibility of ethical scepticism which evolutionistic ethics (as well as intuitive or rationalistic ethics) has overlooked. No demonstration can show that the results of the ethical development are definitive and universal. We meet here again with the important opposition of systematisation and evolution. There will, I think, always be an open question here, though comparative ethics, of which we have so far only the first attempts, can do much to throw light on it.

It would carry us too far to discuss all the philosophical works on ethics, which have been influenced directly or indirectly by evolutionism. I may, however, here refer to the book of C. M. Williams, A Review of the Systems of Ethics founded on the Theory of Evolution,[210] in which, besides Darwin, the following authors are reviewed: Wallace, Haeckel, Spencer, Fiske, Rolph, Barratt, Stephen, Carneri, Hefding, Gizycki, Alexander. As works which criticise evolutionistic ethics from an intuitive point of view and in an instructive way, may be cited: Guyau, La morale anglaise contemporaine,[211] and Sorley, Ethics of Naturalism. I will only mention some interesting contributions to ethical discussion which can be found in Darwinism besides the idea of struggle for life.

The attention which Darwin has directed to variations has opened our eyes to the differences in human nature as well as in nature generally. There is here a fact of great importance for ethical thought, no matter from what ultimate premiss it starts. Only from a very abstract point of view can different individuals be treated in the same manner. The most eminent ethical thinkers, men such as Jeremy Bentham and Immanuel Kant, who discussed ethical questions from very opposite standpoints, agreed in regarding all men as equal in respect of ethical endowment. In regard to Bentham, Leslie Stephen remarks: "He is determined to be thoroughly empirical, to take men as he found them. But his utilitarianism supposed that men's views of happiness and utility were uniform and clear, and that all that was wanted was to show them the means by which their ends could be reached."[212] And Kant supposed that every man would find the "categorical imperative" in his consciousness, when he came to sober reflexion, and that all would have the same qualifications to follow it. But if continual variations, great or small, are going on in human nature, it is the duty of ethics to make allowance for them, both in making claims, and in valuing what is done. A new set of ethical problems have their origin here.[213] It is an interesting fact that Stuart Mill's book On Liberty appeared in the same year as The Origin of Species. Though Mill agreed with Bentham about the original equality of all men's endowments, he regarded individual differences as a necessary result of physical and social influences, and he claimed that free play shall be allowed to differences of character so far as is possible without injury to other men. It is a condition of individual and social progress that a man's mode of action should be determined by his own character and not by tradition and custom, nor by abstract rules. This view was to be corroborated by the theory of Darwin.

But here we have reached a point of view from which the criticism, which in recent years has often been directed against Darwin--that small variations are of no importance in the struggle for life--is of no weight. From an ethical standpoint, and particularly from the ethical standpoint of Darwin himself, it is a duty to foster individual differences that can be valuable, even though they can neither be of service for physical preservation nor be physically inherited. The distinction between variation and mutation is here without importance. It is quite natural that biologists should be particularly interested in such variations as can be inherited and produce new species. But in the human world there is not only a physical, but also a mental and social heredity. When an ideal human character has taken form, then there is shaped a type, which through imitation and influence can become an important factor in subsequent development, even if it cannot form a species in the biological sense of the word. Spiritually strong men often succumb in the physical struggle for life; but they can nevertheless be victorious through the typical influence they exert, perhaps on very distant generations, if the remembrance of them is kept alive, be it in legendary or in historical form. Their very failure can show that a type has taken form which is maintained at all risks, a standard of life which is adhered to in spite of the strongest opposition. The question "to be or not to be" can be put from very different levels of being: it has too often been considered a consequence of Darwinism that this question is only to be put from the lowest level. When a stage is reached, where ideal (ethical, intellectual, aesthetic) interests are concerned, the struggle for life is a struggle for the preservation of this stage. The giving up of a higher standard of life is a sort of death; for there is not only a physical, there is also a spiritual, death.

VI

The Socratic character of Darwin's mind appears in his wariness in drawing the last consequences of his doctrine, in contrast both with the audacious theories of so many of his followers and with the consequences which his antagonists were busy in drawing. Though he, as we have seen, saw from the beginning that his hypothesis would occasion "a whole of metaphysics," he was himself very reserved as to the ultimate questions, and his answers to such questions were extorted from him.

As to the question of optimism and pessimism, Darwin held that though pain and suffering were very often the ways by which animals were led to pursue that course of action which is most beneficial to the species, yet pleasurable feelings were the most habitual guides. "We see this in the pleasure from exertion, even occasionally from great exertion of the body or mind, in the pleasure of our daily meals, and especially in the pleasure derived from sociability, and from loving our families." But there was to him so much suffering in the world that it was a strong argument against the existence of an intelligent First Cause.[214]

It seems to me that Darwin was not so clear on another question, that of the relation between improvement and adaptation. He wrote to Lyell: "When you contrast natural selection and 'improvement,' you seem always to overlook ... that every step in the natural selection of each species implies improvement in that species in relation to its condition of life.... Improvement implies, I suppose, each form obtaining many parts or organs, all excellently adapted for their functions." "All this," he adds, "seems to me quite compatible with certain forms fitted for simple conditions, remaining unaltered, or being, degraded."[215] But the great question is, if the conditions of life will in the long run favour "improvement" in the sense of differentiation (or harmony of differentiation and integration). Many beings are best adapted to their conditions of life if they have few organs and few necessities. Pessimism would not only be the consequence, if suffering outweighed happiness, but also if the most elementary forms of happiness were predominant, or if there were a tendency to reduce the standard of life to the simplest possible, the contentment of inertia or stable equilibrium. There are animals which are very highly differentiated and active in their young state, but later lose their complex organisation and concentrate themselves on the one function of nutrition. In the human world analogies to this sort of adaptation are not wanting. Young "idealists" very often end as old "Philistines." Adaptation and progress are not the same.

Another question of great importance in respect to human evolution is, whether there will be always a possibility for the existence of an impulse to progress, an impulse to make great claims on life, to be active and to alter the conditions of life instead of adapting to them in a passive manner. Many people do not develop because they have too few necessities, and because they have no power to imagine other conditions of life than those under which they live. In his remarks on "the pleasure from exertion" Darwin has a point of contact with the practical idealism of former times--with the ideas of Lessing and Goethe, of Condorcet and Fichte. The continual striving which was the condition of salvation to Faust's soul, is also the condition of salvation to mankind. There is a holy fire which we ought to keep burning, if adaptation is really to be improvement. If, as I have tried to show in my Philosophy of Religion, the innermost core of all religion is faith in the persistence of value in the world, and if the highest values express themselves in the cry "Excelsior!" then the capital point is, that this cry should always be heard and followed. We have here a corollary of the theory of evolution in its application to human life.

Darwin declared himself an agnostic, not only because he could not harmonise the large amount of suffering in the world with the idea of a God as its first cause, but also because he "was aware that if we admit a first cause, the mind still craves to know whence it came and how it arose."[216]

He saw, as Kant had seen before him and expressed in his Kritik der Urtheilskraft, that we cannot accept either of the only two possibilities which we are able to conceive: chance (or brute force) and design. Neither mechanism nor teleology can give an absolute answer to ultimate questions. The universe, and especially the organic life in it, can neither be explained as a mere combination of absolute elements nor as the effect of a constructing thought. Darwin concluded, as Kant, and before him Spinoza, that the oppositions and distinctions which our experience presents, cannot safely be regarded as valid for existence in itself. And, with Kant and Fichte, he found his stronghold in the conviction that man has something to do, even if he cannot solve all enigmas. "The safest conclusion seems to me that the whole subject is beyond the scope of man's intellect; but man can do his duty."[217]

Is this the last word of human thought? Does not the possibility, that man can do his duty, suppose that the conditions of life allow of continuous ethical striving, so that there is a certain harmony between cosmic order and human ideals? Darwin himself has shown how the consciousness of duty can arise as a natural result of evolution. Moreover there are lines of evolution which have their end in ethical idealism, in a kingdom of values, which must struggle for life as all things in the world must do, but a kingdom which has its firm foundation in reality.

FOOTNOTES:

[Footnote 195: Life and Letters of Charles Darwin, Vol. I. p. 8.]

[Footnote 196: Encyclop 銃 ie der philosophischen Wissenschaften (4th edit.), Berlin, 1845, ?249.]

[Footnote 197: Lehrbuch der Naturphilosophie, Jena, 1809.]

[Footnote 198: Ueber den Willen in der Natur (2nd edit.), Frankfurt a. M., 1854, pp. 41-43.]

[Footnote 199: Spencer, Autobiography, Vol. II. p. 50, London and New York, 1904.]

[Footnote 200: Autobiography, Vol. II. p. 100.]

[Footnote 201: Cf. my letter to him 1876, now printed in Duncan's Life and Letters of Herbert Spencer, p. 178. London, 1908.]

[Footnote 202: The present writer, many years ago, in his Psychology (Copenhagen, 1882; Eng. transl. London, 1891), criticised the evolutionistic treatment of the problem of knowledge from the Kantian point of view.]

[Footnote 203: Life and Letters, Vol. II. p. 37.]

[Footnote 204: Ibid. p. 232.]

[Footnote 205: The new science of Ecology occupies an intermediate position between the biography of species and the biography of individuals. Compare Congress of Arts and Science, St. Louis, Vol. V. 1906 (The Reports of Drude and Robinson) and the work of my colleague, E. Warming.]

[Footnote 206: Cf. my History of Modern Philosophy (Eng. transl. London, 1900), I. pp. 76-79.]

[Footnote 207: "Herrschaft und Knechtschaft," Phomenologie des Geistes, IV. A., Leiden, 1907.]

[Footnote 208: The Descent of Man, Vol. I. Ch. iii.]

[Footnote 209: The works of Westermarck and Hobhouse throw new light on many of these features.]

[Footnote 210: New York and London, 1893.]

[Footnote 211: Paris, 1879.]

[Footnote 212: English literature and society in the eighteenth century, London, 1904, p. 187.]

[Footnote 213: Cf. my paper, "The law of relativity in Ethics," International Journal of Ethics, Vol. I. 1891, pp. 37-62.]

[Footnote 214: Life and Letters, Vol. I. p. 310.]

[Footnote 215: Ibid. Vol. II. p. 177.]

[Footnote 216: Life and Letters, Vol. 1. p. 306.]

[Footnote 217: Life and Letters, p. 307.]

VIII

THE INFLUENCE OF DARWIN UPON RELIGIOUS THOUGHT

BY P. N. WAGGETT, M.A., S.S.J.E.

I

The object of this paper is first to point out certain elements of the Darwinian influence upon Religious thought, and then to show reason for the conclusion that it has been, from a Christian point of view, satisfactory. I shall not proceed further to urge that the Christian apologetic in relation to biology has been successful. A variety of opinions may be held on this question, without disturbing the conclusion that the movements of readjustment have been beneficial to those who remain Christians, and this by making them more Christian and not only more liberal. The theologians may sometimes have retreated, but there has been an advance of theology. I know that this account incurs the charge of optimism. It is not the worst that could be made. The influence has been limited in personal range, unequal, even divergent, in operation, and accompanied by the appearance of waste and mischievous products. The estimate which follows requires for due balance a full development of many qualifying considerations. For this I lack space, but I must at least distinguish my view from the popular one that our difficulties about religion and natural science have come to an end.

Concerning the older questions about origins--the origin of the world, of species, of man, of reason, conscience, religion--a large measure of understanding has been reached by some thoughtful men. But meanwhile new questions have arisen, questions about conduct, regarding both the reality of morals and the rule of right action for individuals and societies. And these problems, still far from solution, may also be traced to the influence of Darwin. For they arise from the renewed attention to heredity, brought about by the search for the causes of variation, without which the study of the selection of variations has no sufficient basis.

Even the existing understanding about origins is very far from universal. On these points there were always thoughtful men who denied the necessity of conflict, and there are still thoughtful men who deny the possibility of a truce.

It must further be remembered that the earlier discussion now, as I hope to show, producing favourable results, created also for a time grave damage, not only in the disturbance of faith and the loss of men--a loss not repaired by a change in the currents of debate--but in what I believe to be a still more serious respect. I mean the introduction of a habit of facile and untested hypothesis in religious as in other departments of thought.

Darwin is not responsible for this, but he is in part the cause of it. Great ideas are dangerous guests in narrow minds; and thus it has happened that Darwin--the most patient of scientific workers, in whom hypothesis waited upon research, or if it provisionally outstepped it did so only with the most scrupulously careful acknowledgment--has led smaller and less conscientious men in natural science, in history, and in theology to an over-eager confidence in probable conjecture and a loose grip upon the facts of experience. It is not too much to say that in many quarters the age of materialism was the least matter-of-fact age conceivable, and the age of science the age which showed least of the patient temper of inquiry.

I have indicated, as shortly as I could, some losses and dangers which in a balanced account of Darwin's influence would be discussed at length.

One other loss must be mentioned. It is a defect in our thought which, in some quarters, has by itself almost cancelled all the advantages secured. I mean the exaggerated emphasis on uniformity or continuity; the unwillingness to rest any part of faith or of our practical expectation upon anything that from any point of view can be called exceptional. The high degree of success reached by naturalists in tracing, or reasonably conjecturing, the small beginnings of great differences, has led the inconsiderate to believe that anything may in time become anything else.

It is true that this exaggeration of the belief in uniformity has produced in turn its own perilous reaction. From refusing to believe whatever can be called exceptional, some have come to believe whatever can be called wonderful.

But, on the whole, the discontinuous or highly various character of experience received for many years too little deliberate attention. The conception of uniformity which is a necessity of scientific description has been taken for the substance of history. We have accepted a postulate of scientific method as if it were a conclusion of scientific demonstration. In the name of a generalisation which, however just on the lines of a particular method, is the prize of a difficult exploit of reflexion, we have discarded the direct impressions of experience; or, perhaps it is more true to say, we have used for the criticism of alleged experiences a doctrine of uniformity which is only valid in the region of abstract science. For every science depends for its advance upon limitation of attention, upon the selection out of the whole content of consciousness of that part or aspect which is measurable by the method of the science. Accordingly there is a science of life which rightly displays the unity underlying all its manifestations. But there is another view of life, equally valid, and practically sometimes more important, which recognises the immediate and lasting effect of crisis, difference, and revolution. Our ardour for the demonstration of uniformity of process and of minute continuous change needs to be balanced by a recognition of the catastrophic element in experience, and also by a recognition of the exceptional significance for us of events which may be perfectly regular from an impersonal point of view.

An exorbitant jealousy of miracle, revelation, and ultimate moral distinctions has been imported from evolutionary science into religious thought. And it has been a damaging influence, because it has taken men's attention from facts, and fixed them upon theories.

II

With this acknowledgment of important drawbacks, requiring many words for their proper description, I proceed to indicate certain results of Darwin's doctrine which I believe to be in the long run wholly beneficial to Christian thought. These are:

The encouragement in theology of that evolutionary method of observation and study, which has shaped all modern research:

The recoil of Christian apologetics towards the ground of religious experience, a recoil produced by the pressure of scientific criticism upon other supports of faith:

The restatement, or the recovery of ancient forms of statement, of the doctrines of Creation and of divine Design in Nature, consequent upon the discussion of evolution and of natural selection as its guiding factor.

(1) The first of these is quite possibly the most important of all. It was well defined in a notable paper read by Dr. Gore, now Bishop of Birmingham, to the Church Congress at Shrewsbury in 1896. We have learnt a new caution both in ascribing and in denying significance to items of evidence, in utterance or in event. There has been, as in art, a study of values, which secures perspective and solidity in our representation of facts. On the one hand, a given utterance or event cannot be drawn into evidence as if all items were of equal consequence, like sovereigns in a bag. The question whence and whither must be asked, and the particular thing measured as part of a series. Thus measured it is not less truly important, but it may be important in a lower degree. On the other hand, and for exactly the same reason, nothing that is real is unimportant. The "failures" are not mere mistakes. We see them, in St. Augustine's words, as "scholar's faults which men praise in hope of fruit."

We cannot safely trace the origin of the evolutionistic method to the influence of natural science. The view is tenable that theology led the way. Probably this is a case of alternate and reciprocal debt. Quite certainly the evolutionist method in theology, in Christian history and in the estimate of scripture, has received vast reinforcement from biology, in which evolution has been the ever present and ever victorious conception.

(2) The second effect named is the new willingness of Christian thinkers to take definite account of religious experience. This is related to Darwin through the general pressure upon religious faith of scientific criticism. The

great advance of our knowledge of organisms has been an important element in the general advance of science. It has acted, by the varied requirements of the theory of organisms, upon all other branches of natural inquiry, and it held for a long time that leading place in public attention which is now occupied by speculative physics. Consequently it contributed largely to our present estimation of science as the supreme judge in all matters of inquiry,[218] to the supposed destruction of mystery and the disparagement of metaphysics which marked the last age, as well as to the just recommendation of scientific method in branches of learning where the direct acquisitions of natural science had no place.

Besides this, the new application of the idea of law and mechanical regularity to the organic world seemed to rob faith of a kind of refuge. The romantics had, as Berthelot[219] shows, appealed to life to redress the judgments drawn from mechanism. Now, in Spencer, evolution gave us a vitalist mechanic or mechanical vitalism, and the appeal seemed cut off. We may return to this point later when we consider evolution; at present I only endeavour to indicate that general pressure of scientific criticism which drove men of faith to seek the grounds of reassurance in a science of their own; in a method of experiment, of observation, of hypothesis checked by known facts. It is impossible for me to do more than glance across the threshold of this subject. But it is necessary to say that the method is in an elementary stage of revival. The imposing success that belongs to natural science is absent: we fall short of the unchallengeable unanimity of the Biologists on fundamentals. The experimental method with its sure repetitions cannot be applied to our subject-matter. But we have something like the observational method of palaeontology and geographical distribution; and in biology there are still men who think that the large examination of varieties by way of geography and the search of strata is as truly scientific, uses as genuinely the logical method of difference, and is as fruitful in sure conclusions as the quasi-chemical analysis of Mendelian laboratory work, of which last I desire to express my humble admiration. Religion also has its observational work in the larger and possibly more arduous manner.

But the scientific work in religion makes its way through difficulties and dangers. We are far from having found the formula of its combination with the historical elements of our apologetic. It is exposed, therefore, to a damaging fire not only from unspiritualist psychology and pathology but also from the side of scholastic dogma. It is hard to admit on equal terms a partner to the old undivided rule of books and learning. With Charles Lamb, we cry in some distress, "must knowledge come to me, if it come at all, by some awkward experiment of intuition, and no longer by this familiar process of reading?"[220] and we are answered that the old process has an imperishable value, only we have not yet made clear its connection with other contributions. And all the work is young, liable to be drawn into unprofitable excursions, side-tracked by self-deceit and pretence; and it fatally attracts, like the older mysticism, the curiosity and the expository powers of those least in sympathy with it, ready writers who, with all the air of extended research, have been content with narrow grounds for induction. There is a danger, besides, which accompanies even the most genuine work of this science and must be provided against by all its serious students. I mean the danger of unbalanced introspection both for individuals and for societies; of a preoccupation comparable to our modern social preoccupation with bodily health; of reflexion upon mental states not accompanied by exercise and growth of the mental powers; the danger of contemplating will and neglecting work, of analysing conviction and not criticising evidence.

Still, in spite of dangers and mistakes, the work remains full of hopeful indications, and, in the best examples,[221] it is truly scientific in its determination to know the very truth, to tell what we think, not what we think we ought to think,[222] truly scientific in its employment of hypothesis and verification, and in growing conviction of the reality of its subject-matter through the repeated victories of a mastery which advances, like science, in the Baconian road of obedience. It is reasonable to hope that progress in this respect will be more rapid and sure when religious study enlists more men affected by scientific desire and endowed with scientific capacity.

The class of investigating minds is a small one, possibly even smaller than that of reflecting minds. Very few persons at any period are able to find out anything whatever. There are few observers, few discoverers, few who even wish to discover truth. In how many societies the problems of philology which face every person who speaks English are left unattempted! And if the inquiring or the successfully inquiring class of minds is small, much smaller, of course, is the class of those possessing the scientific aptitude in an eminent degree. During the last age this most distinguished class was to a very great extent absorbed in the study of phenomena, a study which had fallen into arrears. For we stood possessed, in rudiment, of means of observation, means for travelling and acquisition, qualifying men for a larger knowledge than had yet been attempted. These were now to be directed with new accuracy and ardour upon the fabric and behaviour of the world of sense. Our debt to the great masters in physical science who overtook and almost outstripped the task cannot be measured; and, under the honourable leadership of Ruskin, we may all well do penance if we have failed "in the respect due to their great powers of thought, or in the admiration due to the far scope of their discovery."[223] With what miraculous mental energy and divine good fortune--as Romans said of their soldiers--did our men of curiosity face the apparently impenetrable mysteries of nature! And how natural it was that immense accessions of knowledge, unrelated to the spiritual facts of life, should discredit Christian faith, by the apparent superiority of the new work to the feeble and unprogressive knowledge of Christian believers! The day is coming when men of this mental character and rank, of this curiosity, this energy and this good fortune in investigation, will be employed in opening mysteries of a spiritual nature. They will silence with masterful witness the over-confident denials of naturalism. They will be in danger of the widespread recognition which thirty years ago accompanied every utterance of Huxley, Tyndall, Spencer. They will contribute, in spite of adulation, to the advance of sober religious and moral science.

And this result will be due to Darwin, first because by raising the dignity of natural science, he encouraged the development of the scientific mind; secondly because he gave to religious students the example of patient and ardent investigation; and thirdly because by the pressure of naturalistic criticism the religious have been driven to ascertain the causes of their own convictions, a work in which they were not without the sympathy of men of science.[224]

In leaving the subject of scientific religious inquiry, I will only add that I do not believe it receives any important help--and certainly it suffers incidentally much damaging interruption--from the study of abnormal manifestations or abnormal conditions of personality.

(3) Both of the above effects seem to me of high, perhaps the very highest, importance to faith and to thought. But, under the third head, I name two which are more directly traceable to the personal work of Darwin, and more

definitely characteristic of the age in which his influence was paramount: viz. the influence of the two conceptions of evolution and natural selection upon the doctrine of creation and of design respectively.

It is impossible here, though it is necessary for a complete sketch of the matter, to distinguish the different elements and channels of this Darwinian influence; in Darwin's own writings, in the vigourous polemic of Huxley, and strangely enough, but very actually for popular thought, in the teaching of the definitely anti-Darwinian evolutionist Spencer.

Under the head of the directly and purely Darwinian elements I should class as preeminent the work of Wallace and of Bates; for no two sets of facts have done more to fix in ordinary intelligent minds a belief in organic evolution and in natural selection as its guiding factor than the facts of geographical distribution and of protective colour and mimicry. The facts of geology were difficult to grasp and the public and theologians heard more often of the imperfection than of the extent of the geological record. The witness of embryology, depending to a great extent upon microscopic work, was and is beyond the appreciation of persons occupied in fields of work other than biology.

III

From the influence in religion of scientific modes of thought we pass to the influence of particular biological conceptions. The former effect comes by way of analogy, example, encouragement and challenge; inspiring or provoking kindred or similar modes of thought in the field of theology; the latter by a collision of opinions upon matters of fact or conjecture which seem to concern both science and religion.

In the case of Darwinism the story of this collision is familiar, and falls under the heads of evolution and natural selection, the doctrine of descent with modification, and the doctrine of its guidance or determination by the struggle for existence between related varieties. These doctrines, though associated and interdependent, and in popular thought not only combined but confused, must be considered separately. It is true that the ancient doctrine of Evolution, in spite of the ingenuity and ardour of Lamarck, remained a dream tantalising the intellectual ambition of naturalists, until the day when Darwin made it conceivable by suggesting the machinery of its guidance. And, further, the idea of natural selection has so effectively opened the door of research and stimulated observation in a score of principal directions that, even if the Darwinian explanation became one day much less convincing than, in spite of recent criticism, it now is, yet its passing, supposing it to pass, would leave the doctrine of Evolution immeasurably and permanently strengthened. For in the interests of the theory of selection, "Darwin," as Muller wrote, facts have been collected which remain in any case evidence of the reality of descent with modification.

But still, though thus united in the modern history of convictions, though united and confused in the collision of biological and traditional opinion, yet evolution and natural selection must be separated in theological no less than in biological estimation. Evolution seemed inconsistent with Creation; natural selection with Providence and Divine design.

Discussion was maintained about these points for many years and with much dark heat. It ranged over many particular topics and engaged minds different in tone, in quality, and in accomplishment. There was at most times a degree of misconception. Some naturalists attributed to theologians in general a poverty of thought which belonged really to men of a particular temper or training. The "timid theism" discerned in Darwin by so cautious a theologian as Liddon[225] was supposed by many biologists to be the necessary foundation of an honest Christianity. It was really more characteristic of devout naturalists like Philip Henry Gosse, than of religious believers as such.[226] The study of theologians more considerable and even more typically conservative than Liddon does not confirm the description of religious intolerance given in good faith, but in serious ignorance, by a disputant so acute, so observant and so candid as Huxley. Something hid from each other's knowledge the devoted pilgrims in two great ways of thought. The truth may be, that naturalists took their view of what creation was from Christian men of science who naturally looked in their own special studies for the supports and illustrations of their religious belief. Of almost every labourious student it may be said: "Hic ab arte sua non recessit." And both the believing and the denying naturalists, confining habitual attention to a part of experience, are apt to affirm and deny with trenchant vigour and something of a narrow clearness "Qui respiciunt ad pauca, de facili pronunciant."[227]

Newman says of some secular teachers that "they persuade the world of what is false by urging upon it what is true." Of some early opponents of Darwin it might be said by a candid friend that, in all sincerity of devotion to truth, they tried to persuade the world of what is true by urging upon it what is false. If naturalists took their version of orthodoxy from amateurs in theology, some conservative Christians, instead of learning what evolution meant to its regular exponents, took their view of it from celebrated persons, not of the front rank in theology or in thought, but eager to take account of public movements and able to arrest public attention.

Cleverness and eloquence on both sides certainly had their share in producing the very great and general disturbance of men's minds in the early days of Darwinian teaching. But by far the greater part of that disturbance was due to the practical novelty and the profound importance of the teaching itself, and to the fact that the controversy about evolution quickly became much more public than any controversy of equal seriousness had been for many generations.

We must not think lightly of that great disturbance because it has, in some real sense, done its work, and because it is impossible in days of more coolness and light, to recover a full sense of its very real difficulties.

Those who would know them better should add to the calm records of Darwin[228] and to the story of Huxley's impassioned championship, all that they can learn of George Romanes.[229] For his life was absorbed in this very struggle and reproduced its stages. It began in a certain assured simplicity of biblical interpretation; it went on, through the glories and adventures of a paladin in Darwin's train, to the darkness and dismay of a man who saw all his most cherished beliefs rendered, as he thought, incredible.[230] He lived to find the freer faith for which process and purpose are not irreconcilable, but necessary to one another. His development, scientific, intellectual and moral, was itself of high significance; and its record is of unique value to our own generation, so near the age of that doubt and yet so far from it; certainly still much in need of the caution and courage by which past endurance prepares men for new emergencies. We have little enough reason to be sure that in the discussions awaiting us we shall do as well as our predecessors in theirs. Remembering their endurance of mental pain, their ardour in mental labour, the heroic temper and the high sincerity of controversialists on either side, we may well speak of our fathers in such words of modesty and self-judgment as Drayton used when he sang the victors of Agincourt. The progress of biblical study, in the departments of Introduction and Exegesis, resulting in the recovery of a point of view anciently tolerated if not prevalent, has altered some of the conditions of that discussion. In the years near 1858, the witness of Scripture was adduced both by Christian advocates and their critics as if unmistakably irreconcilable with Evolution.

Huxley[231] found the path of the blameless naturalist everywhere blocked by "Moses": the believer in revelation was generally held to be forced to a choice between revealed cosmogony and the scientific account of origins. It is not clear how far the change in Biblical interpretation is due to natural science, and how far to the vital movements of theological study which have been quite independent of the controversy about species. It belongs to a general renewal of Christian movement, the recovery of a heritage. "Special Creation"--really a biological rather than a theological conception,--seems in its rigid form to have been a recent element even in English biblical orthodoxy.

The Middle Ages had no suspicion that religious faith forbad inquiry into the natural origination of the different forms of life. Bartholomaeus Anglicus, an English Franciscan of the thirteenth century, was a mutationist in his way, as Aristotle, "the Philosopher" of the Christian Schoolmen, had been in his. So late as the seventeenth century, as we learn not only from early proceedings of the Royal Society, but from a writer so homely and so regularly pious as Walton, the variation of species and "spontaneous" generations had no theological bearing, except as instances of that various wonder of the world which in devout minds is food for devotion.

It was in the eighteenth century that the harder statement took shape. Something in the preciseness of that age, its exaltation of law, its cold passion for a stable and measured universe, its cold denial, its cold affirmation of the power of God, a God of ice, is the occasion of that rigidity of religious thought about the living world which Darwin by accident challenged, or rather by one of those movements of genius which, Goethe[232] declares, are "elevated above all earthly control."

If religious thought in the eighteenth century was aimed at a fixed and nearly finite world of spirit, it followed in all these respects the secular and critical lead. In fact, religion in a mechanical age is condemned if it takes any but a mechanical tone. Butler's thought was too moving, too vital, too evolutionary, for the sceptics of his time. In a rationalist, encyclopaedic period, religion also must give hard outline to its facts, it must be able to display its secret to any sensible man in the language used by all sensible men. Milton's prophetic genius furnished the eighteenth century, out of the depth of the passionate age before it, with the theological tone it was to need. In spite of the austere magnificence of his devotion, he gives to smaller souls a dangerous lead. The rigidity of Scripture exegesis belonged to this stately but imperfectly sensitive mode of thought. It passed away with the influence of the older rationalists whose precise denials matched the precise and limited affirmations of the static orthodoxy.

I shall, then, leave the specially biblical aspect of the debate--interesting as it is and even useful, as in Huxley's correspondence with the Duke of Argyll and others in 1892[234]--in order to consider without complication the permanent elements of Christian thought brought into question by the teaching of evolution.

Such permanent elements are the doctrine of God as Creator of the universe, and the doctrine of man as spiritual and unique. Upon both the doctrine of evolution seemed to fall with crushing force.

With regard to Man I leave out, acknowledging a grave omission, the doctrine of the Fall and of Sin. And I do so because these have not yet, as I believe, been adequately treated: here the fruitful reaction to the stimulus of evolution is yet to come. The doctrine of sin, indeed, falls principally within the scope of that discussion which has followed or displaced the Darwinian; and without it the Fall cannot be usefully considered. For the question about the Fall is a question not merely of origins, but of the interpretation of moral facts whose moral reality must first be established.

I confine myself therefore to Creation and the dignity of man.

The meaning of evolution, in the most general terms, is that the differentiation of forms is not essentially separate from their behaviour and use; that if these are within the scope of study, that is also; that the world has taken the form we see by movements not unlike those we now see in progress; that what may be called proximate origins are continuous in the way of force and matter, continuous in the way of life, with actual occurrences and actual characteristics. All this has no revolutionary bearing upon the question of ultimate origins. The whole is a statement about process. It says nothing to metaphysicians about cause. It simply brings within the scope of observation or conjecture that series of changes which has given their special characters to the different parts of the world we see. In particular, evolutionary science aspires to the discovery of the process or order of the appearance of life itself: if it were to achieve its aim it could say nothing of the cause of this or indeed of the most familiar occurrences. We should have become spectators or convinced historians of an event which, in respect of its cause and ultimate meaning, would be still impenetrable.

With regard to the origin of species, supposing life already established, biological science has the well founded hopes and the measure of success with which we are all familiar. All this has, it would seem, little chance of collision with a consistent theism, a doctrine which has its own difficulties unconnected with any particular view of order or process. But when it was stated that species had arisen by processes through which new species were still being made, evolutionism came into collision with a statement, traditionally religious, that species were formed and fixed once for all and long ago.

What is the theological import of such a statement when it is regarded as essential to belief in God? Simply that God's activity, with respect to the formation of living creatures, ceased at some point in past time.

"God rested" is made the touchstone of orthodoxy. And when, under the pressure of the evidences, we found ourselves obliged to acknowledge and assert the present and persistent power of God, in the maintenance and in the continued formation of "types," what happened was the abolition of a time-limit. We were forced only to a bolder claim, to a theistic language less halting, more consistent, more thorough in its own line, as well as better qualified to assimilate and modify such schemes as Von Hartmann's philosophy of the Unconscious--a philosophy, by the way, quite intolerant of a merely mechanical evolution.[235]

Here was not the retrenchment of an extravagant assertion, but the expansion of one which was faltering and inadequate. The traditional statement did not need paring down so as to pass the meshes of a new and exacting criticism. It was itself a net meant to surround and enclose experience; and we must increase its size and close its mesh to hold newly disclosed facts of life. The world, which had seemed a fixed picture or model, gained first perspective and then solidity and movement. We had a glimpse of organic history; and Christian thought became more living and more assured as it met the larger view of life.

However unsatisfactory the new attitude might be to our critics, to Christians the reform was positive. What was discarded was a limitation, a negation. The movement was essentially conservative, even actually reconstructive. For the language disused was a language inconsistent with the definitions of orthodoxy; it set bounds to the infinite, and by implication withdrew from the creative rule all such processes as could be brought within the descriptions of research. It ascribed fixity and finality to that "creature" in which an apostle taught us to recognise the birth-struggles of an unexhausted progress. It tended to banish mystery from the world we see, and to confine it to a remote first age.

In the reformed, the restored, language of religion, Creation became again not a link in a rational series to complete a circle of the sciences, but the mysterious and permanent relation between the infinite and the finite, between the moving changes we know in part, and the Power, after the fashion of that observation, unknown, which is itself "unmoved all motion's source."[236]

With regard to man it is hardly necessary, even were it possible, to illustrate the application of this bolder faith. When the record of his high extraction fell under dispute, we were driven to a contemplation of the whole of his life, rather than of a part and that part out of sight. We remembered again, out of Aristotle, that the result of a process interprets its beginnings. We were obliged to read the title of such dignity as we may claim, in results and still more in aspirations.

Some men still measure the value of great present facts in life--reason and virtue and sacrifice--by what a self-disparaged reason can collect of the meaner rudiments of these noble gifts. Mr. Balfour has admirably displayed the discrepancy, in this view, between the alleged origin and the alleged authority of reason. Such an argument ought to be used not to discredit the confident reason, but to illuminate and dignify its dark beginnings, and to show that at every step in the long course of growth a Power was at work which is not included in any term or in all the terms of the series.

I submit that the more men know of actual Christian teaching, its fidelity to the past, and its sincerity in face of discovery, the more certainly they will judge that the stimulus of the doctrine of evolution has produced in the long run vigour as well as flexibility in the doctrine of Creation and of man.

I pass from Evolution in general to Natural Selection.

The character in religious language which I have for short called mechanical was not absent in the argument from design as stated before Darwin. It seemed to have reference to a world conceived as fixed. It pointed, not to the plastic capacity and energy of living matter, but to the fixed adaptation of this and that organ to an unchanging place or function.

Mr. Hobhouse has given us the valuable phrase "a niche of organic opportunity." Such a phrase would have borne a different sense in non-evolutionary thought. In that thought, the opportunity was an opportunity for the Creative Power, and Design appeared in the preparation of the organism to fit the niche. The idea of the niche and its occupant growing together from simpler to more complex mutual adjustment was unwelcome to this teleology. If the adaptation was traced to the influence, through competition, of the environment, the old teleology lost an illustration and a proof. For the cogency of the proof in every instance depended upon the absence of explanation. Where the process of adaptation was discerned, the evidence of Purpose or Design was weak. It was strong only when the natural antecedents were not discovered, strongest when they could be declared undiscoverable.

Paley's favourite word is "Contrivance"; and for him contrivance is most certain where production is most obscure. He points out the physiological advantage of the valvulae conniventes to man, and the advantage for teleology of the fact that they cannot have been formed by "action and pressure." What is not due to pressure may be attributed to design, and when a "mechanical" process more subtle than pressure was suggested, the case for design was so far weakened. The cumulative proof from the multitude of instances began to disappear when, in selection, a natural sequence was suggested in which all the adaptations might be reached by the motive power of life, and especially when, as in Darwin's teaching, there was full recognition of the reactions of life to the stimulus of circumstance. "The organism fits the niche," said the teleologist, "because the Creator formed it so as to fit." "The organism fits the niche," said the naturalist, "because unless it fitted it could not exist." "It was fitted to survive," said the theologian. "It survives because it fits," said the selectionist. The two forms of statement are not incompatible; but the new statement, by provision of an ideally universal explanation of process, was hostile to a doctrine of purpose which relied upon evidences always exceptional however numerous. Science persistently presses on to find the universal machinery of adaptation in this planet; and whether this be found in selection, or in direct-effect, or in vital reactions resulting in large changes, or in a combination of these and other factors, it must always be opposed to the conception of a Divine Power here and there but not everywhere active.

For science, the Divine must be constant, operative everywhere and in every quality and power, in environment and in organism, in stimulus and in reaction, in variation and in struggle, in hereditary equilibrium, and in "the unstable state of species"; equally present on both sides of every strain, in all pressures and in all resistances, in short in the general wonder of life and the world. And this is exactly what the Divine Power must be for religious faith.

The point I wish once more to make is that the necessary readjustment of teleology, so as to make it depend upon the contemplation of the whole instead of a part, is advantageous quite as much to theology as to science. For the older view failed in courage. Here again our theism was not sufficiently theistic.

Where results seemed inevitable, it dared not claim them as God-given. In the argument from Design it spoke not of God in the sense of theology, but of a Contriver, immensely, not infinitely wise and good, working within a world, the scene, rather than the ever dependent outcome, of His Wisdom; working in such emergencies and opportunities as occurred, by forces not altogether within His control, towards an end beyond Himself. It gave us, instead of the awful reverence due to the Cause of all substance and form, all love and wisdom, a dangerously detached appreciation of an ingenuity and benevolence meritorious in aim and often surprisingly successful in contrivance.

The old teleology was more useful to science than to religion, and the design-naturalists ought to be gratefully remembered by Biologists. Their search for evidences led them to an eager study of adaptations and of minute forms, a study such as we have now an incentive to in the theory of Natural Selection. One hardly meets with the same ardour in microscopical research until we come to modern workers. But the argument from Design was never of great importance to faith. Still, to rid it of this character was worth all the stress and anxiety of the gallant old war. If Darwin had done nothing else for us, we are to-day deeply in his debt for this. The world is not less venerable to us now, not less eloquent of the causing mind, rather much more eloquent and sacred. But our wonder is not that "the underjaw of the swine works under the ground" or in any or all of those particular adaptations which Paley collected with so much skill, but that a purpose transcending, though resembling, our own purposes, is everywhere manifest; that what we live in is a whole, mutually sustaining, eventful and beautiful, where the "dead" forces feed the energies of life, and life sustains a stranger existence, able in some real measure to contemplate the whole, of which, mechanically considered, it is a minor product and a rare ingredient. Here, again, the change was altogether positive. It was not the escape of a vessel in a storm with loss of spars and rigging, not a shortening of sail to save the masts and make a port of refuge. It was rather the emergence from narrow channels to an open sea. We had propelled the great ship, finding purchase here and there for slow and uncertain movement. Now, in deep water, we spread large canvas to a favouring breeze.

The scattered traces of design might be forgotten or obliterated. But the broad impression of Order became plainer when seen at due distance and in sufficient range of effect, and the evidence of love and wisdom in the universe could be trusted more securely for the loss of the particular calculation of their machinery.

Many other topics of faith are affected by modern biology. In some of these we have learnt at present only a wise caution, a wise uncertainty. We stand before the newly unfolded spectacle of suffering, silenced; with faith not scientifically reassured but still holding fast certain other clues of conviction. In many important topics we are at a loss. But in others, and among them those I have mentioned, we have passed beyond this negative state and find faith positively strengthened and more fully expressed.

We have gained also a language and a habit of thought more fit for the great and dark problems that remain, less liable to damaging conflicts, equipped for more rapid assimilation of knowledge. And by this change biology itself is a gainer. For, relieved of fruitless encounters with popular religion, it may advance with surer aim along the path of really scientific life-study which was reopened for modern men by the publication of The Origin of Species.

Charles Darwin regretted that, in following science, he had not done "more direct good"[237] to his fellow-creatures. He has, in fact, rendered substantial service to interests bound up with the daily conduct and hopes of common men; for his work has led to improvements in the preaching of the Christian faith.

FOOTNOTES:

[Footnote 218: F. R. Tennant: "The Being of God in the light of Physical Science," in Essays on some theological questions of the day. London, 1905.]

[Footnote 219: Evolutionisme et Platonisme, pp. 45, 46, 47. Paris, 1908.]

[Footnote 220: Essays of Elia, "New Year's Eve," p. 41; Ainger's edition. London, 1899.]

[Footnote 221: Such an example is given in Baron F. von Hotel's recently finished book, the result of thirty years' research: The Mystical Element of Religion, as studied in Saint Catherine of Genoa and her Friends. London, 1908.]

[Footnote 222: G. Tyrrell, in Mediaevalism, has a chapter which is full of the important moral element in a scientific attitude. "The only infallible guardian of truth is the spirit of truthfulness." Mediaevalism, p. 182, London, 1908.]

[Footnote 223: Queen of the Air, Preface, p. vii. London, 1906.]

[Footnote 224: The scientific rank of its writer justifies the insertion of the following letter from the late Sir John Burdon-Sanderson to me. In the lecture referred to I had described the methods of Professor Moseley in teaching Biology as affording a suggestion of the scientific treatment of religion.

OXFORD,

April 30, 1902.

DEAR SIR:

I feel that I must express to you my thanks for the discourse which I had the pleasure of listening to yesterday afternoon.

I do not mean to say that I was able to follow all that you said as to the identity of Method in the two fields of Science and Religion, but I recognise that the "mysticism" of which you spoke gives us the only way by which the two fields can be brought into relation.

Among much that was memorable, nothing interested me more than what you said of Moseley.

No one, I am sure, knew better than you the value of his teaching and in what that value consisted.

Yours faithfully,

J. BURDON-SANDERSON.

[Footnote 225: H. P. Liddon, The Recovery of S. Thomas; a sermon preached in St. Paul's, London, on April 23rd, 1882 (the Sunday after Darwin's death).]

[Footnote 226: Dr. Pusey (Unscience not Science adverse to Faith, 1878) writes: "The questions as to 'species,' of what variations the animal world is capable, whether the species be more or fewer, whether accidental variations may become hereditary ... and the like, naturally fall under the province of science. In all these questions Mr. Darwin's careful observations gained for him a deserved approbation and confidence."]

[Footnote 227: Aristotle, in Bacon, quoted by Newman in his Idea of a University, p. 78. London, 1873.]

[Footnote 228: Life and Letters and More Letters of Charles Darwin.]

[Footnote 229: Life and Letters, London, 1896. Thoughts on Religion, London, 1895. Candid Examination of Theism, London, 1878.]

[Footnote 230: "Never in the history of man has so terrific a calamity befallen the race as that which all who look may now (viz. in consequence of the scientific victory of Darwin) behold advancing as a deluge black with destruction, resistless in might, uprooting our most cherished hopes, engulphing our most precious creed, and burying our highest life in mindless destruction."--A Candid Examination of Theism, p. 51.]

[Footnote 231: Science and Christian Tradition. London, 1904.]

[Footnote 232: "No productiveness of the highest kind ... is in the power of anyone."--Conversations of Goethe with Eckermann and Soret. London, 1850.]

[Footnote 233: Berthelot, Evolutionisme et Platonisme, Paris, 1908, p. 45.]

[Footnote 234: Times, 1892, passim.]

[Footnote 235: See Von Hartmann's Wahrheit und Irrthum in Darwinismus. Berlin, 1875.]

[Footnote 236: Hymn of the Church--

Rerum Deus tenax vigor, Immotus in te permanens.]

[Footnote 237: Life and Letters, Vol. III. p. 359.]

IX

DARWINISM AND HISTORY

BY J. B. BURY, LITT.D., LL.D.

Regius Professor of Modern History in the University of Cambridge

1. Evolution, and the principles associated with the Darwinian theory, could not fail to exert a considerable influence on the studies connected with the history of civilised man. The speculations which are known as "philosophy of history," as well as the sciences of anthropology, ethnography, and sociology (sciences which though they stand on their own feet are for the historian auxiliary), have been deeply affected by these principles. Historiographers, indeed, have with few exceptions made little attempt to apply them; but the growth of historical study in the nineteenth century has been determined and characterised by the same general principle which has underlain the simultaneous developments of the study of nature, namely the genetic idea. The "historical" conception of nature, which has produced the history of the solar system, the story of the earth, the genealogies of telluric organisms, and has revolutionised natural science, belongs to the same order of thought as the conception of human history as a continuous, genetic, causal process--a conception which has revolutionised historical research and made it scientific. Before proceeding to consider the application of evolutional principles, it will be pertinent to notice the rise of this new view.

2. With the Greeks and Romans history had been either a descriptive record or had been written in practical interests. The most eminent of the ancient historians were pragmatical; that is, they regarded history as an instructress in statesmanship, or in the art of war, or in morals. Their records reached back such a short way, their experience was so brief, that they never attained to the conception of continuous process, or realised the significance of time; and they never viewed the history of human societies as a phenomenon to be investigated for its own sake. In the middle ages there was still less chance of the emergence of the ideas of progress and development. Such notions were excluded by the fundamental doctrines of the dominant religion which bounded and bound men's minds. As the course of history was held to be determined from hour to hour by the arbitrary will of an extra cosmic person,

there could be no self-contained causal development, only a dispensation imposed from without. And as it was believed that the world was within no great distance from the end of this dispensation, there was no motive to take much interest in understanding the temporal, which was to be only temporary.

The intellectual movements of the fifteenth and sixteenth centuries prepared the way for a new conception, but it did not emerge immediately. The historians of the Renaissance period simply reverted to the ancient pragmatical view. For Machiavelli, exactly as for Thucydides and Polybius, the use of studying history was instruction in the art of politics. The Renaissance itself was the appearance of a new culture, different from anything that had gone before; but at the time men were not conscious of this; they saw clearly that the traditions of classical antiquity had been lost for a long period, and they were seeking to revive them, but otherwise they did not perceive that the world had moved, and that their own spirit, culture, and conditions were entirely unlike those of the thirteenth century. It was hardly till the seventeenth century that the presence of a new age, as different from the middle ages as from the ages of Greece and Rome, was fully realised. It was then that the triple division of ancient, medieval, and modern was first applied to the history of western civilisation. Whatever objections may be urged against this division, which has now become almost a category of thought, it marks a most significant advance in man's view of his own past. He has become conscious of the immense changes in civilisation which have come about slowly in the course of time, and history confronts him with a new aspect. He has to explain how those changes have been produced, how the transformations were effected. The appearance of this problem was almost simultaneous with the rise of rationalism, and the great historians and thinkers of the eighteenth century, such as Montesquieu, Voltaire, Gibbon, attempted to explain the movement of civilisation by purely natural causes. These brilliant writers prepared the way for the genetic history of the following century. But in the spirit of the Aufklung, that eighteenth-century Enlightenment to which they belonged, they were concerned to judge all phenomena before the tribunal of reason; and the apotheosis of "reason" tended to foster a certain superior a priori attitude, which was not favourable to objective treatment and was incompatible with a "historical sense." Moreover the traditions of pragmatical historiography had by no means disappeared.

3. In the first quarter of the nineteenth century the meaning of genetic history was fully realised. "Genetic" perhaps is as good a word as can be found for the conception which in this century was applied to so many branches of knowledge in the spheres both of nature and of mind. It does not commit us to the doctrine proper of evolution, nor yet to any teleological hypothesis such as is implied in "progress." For history it meant that the present condition of the human race is simply and strictly the result of a causal series (or set of causal series)--a continuous succession of changes, where each state arises causally out of the preceding; and that the business of historians is to trace this genetic process, to explain each change, and ultimately to grasp' the complete development of the life of humanity. Three influential writers, who appeared at this stage and helped to initiate a new period of research, may specially be mentioned. Ranke in 1824 definitely repudiated the pragmatical view which ascribes to history the duties of an instructress, and with no less decision renounced the function, assumed by the historians of the Aufkl 鋌 ung, to judge the past; it was his business, he said, merely to show how things really happened. Niebuhr was already working in the same spirit and did more than any other writer to establish the principle that historical transactions must be related to the ideas and conditions of their age. Savigny about the same time founded the "historical school" of law. He sought to show that law was not the creation of an enlightened will, but grew out of custom and was developed by a series of adaptations and rejections, thus applying the conception of evolution. He helped to diffuse the notion that all the institutions of a society or a nation are as closely interconnected as the parts of a living organism.

4. The conception of the history of man as a causal development meant the elevation of historical inquiry to the dignity of a science. Just as the study of bees cannot become scientific so long as the student's interest in them is only to procure honey or to derive moral lessons from the labours of "the little busy bee," so the history of human societies cannot become the object of pure scientific investigation so long as man estimates its value in pragmatical scales. Nor can it become a science until it is conceived as lying entirely within a sphere in which the law of cause and effect has unreserved and unrestricted dominion. On the other hand, once history is envisaged as a causal process, which contains within itself the explanation of the development of man from his primitive state to the point which he has

reached, such a process necessarily becomes the object of scientific investigation and the interest in it is scientific curiosity.

At the same time, the instruments were sharpened and refined. Here Wolf, a philologist with historical instinct, was a pioneer. His Prolegomena to Homer (1795) announced new modes of attack. Historical investigation was soon transformed by the elaboration of new methods.

5. "Progress" involves a judgment of value, which is not involved in the conception of history as a genetic process. It is also an idea distinct from that of evolution. Nevertheless it is closely related to the ideas which revolutionised history at the beginning of the last century; it swam into men's ken simultaneously; and it helped effectively to establish the notion of history as a continuous process and to emphasise the significance of time. Passing over earlier anticipations, I may point to a Discours of Turgot (1750), where history is presented as a process in which "the total mass of the human race" "marches continually though sometimes slowly to an ever increasing perfection." That is a clear statement of the conception which Turgot's friend Condorcet elaborated in the famous work, published in 1795, Esquisse d'un tableau historique des progre de l'esprit humain. This work first treated with explicit fulness the idea to which a leading role was to fall in the ideology of the nineteenth century. Condorcet's book reflects the triumphs of the Tiers, whose growing importance had also inspired Turgot; it was the political changes in the eighteenth century which led to the doctrine, emphatically formulated by Condorcet, that the masses are the most important element in the historical process. I dwell on this because, though Condorcet had no idea of evolution, the predominant importance of the masses was the assumption which made it possible to apply evolutional principles to history. And it enabled Condorcet himself to maintain that the history of civilisation, a progress still far from being complete, was a development conditioned by general laws.

6. The assimilation of society to an organism, which was a governing notion in the school of Savigny, and the conception of progress, combined to produce the idea of an organic development, in which the historian has to determine the central principle or leading character. This is illustrated by the apotheosis of democracy in Tocqueville's Democratie en Amerique, where the theory is maintained that "the gradual and progressive development of

equality is at once the past and the future of the history of men." The same two principles are combined in the doctrine of Spencer (who held that society is an organism, though he also contemplated its being what he calls a "super-organic aggregate"),[238] that social evolution is a progressive change from militarism to industrialism.

7. The idea of development assumed another form in the speculations of German idealism. Hegel conceived the successive periods of history as corresponding to the ascending phases or ideas in the self-evolution of his Absolute Being. His Lectures on the Philosophy of History were published in 1837 after his death. His philosophy had a considerable effect, direct and indirect, on the treatment of history by historians, and although he was superficial and unscientific himself in dealing with historical phenomena, he contributed much towards making the idea of historical development familiar. Ranke was influenced, if not by Hegel himself, at least by the Idealistic philosophies of which Hegel's was the greatest. He was inclined to conceive the stages in the process of history as marked by incarnations, as it were, of ideas, and sometimes speaks as if the ideas were independent forces, with hands and feet. But while Hegel determined his ideas by a priori logic, Ranke obtained his by induction--by a strict investigation of the phenomena; so that he was scientific in his method and work, and was influenced by Hegelian prepossessions only in the kind of significance which he was disposed to ascribe to his results. It is to be noted that the theory of Hegel implied a judgment of value; the movement was a progress towards perfection.

8. In France, Comte approached the subject from a different side, and exercised, outside Germany, a far wider influence than Hegel. The 4th volume of his Cours de philosophie positive, which appeared in 1839, created sociology and treated history as a part of this new science, namely as "social dynamics." Comte sought the key for unfolding historical development, in what he called the social-psychological point of view, and he worked out the two ideas which had been enunciated by Condorcet: that the historian's attention should be directed not, as hitherto, principally to eminent individuals, but to the collective behaviour of the masses, as being the most important element in the process; and that, as in nature, so in history, there are general laws, necessary and constant, which condition the development. The two points are intimately connected, for it is only when the masses are moved into the foreground that regularity, uniformity, and law can be

conceived as applicable. To determine the social-psychological laws which have controlled the development is, according to Comte, the task of sociologists and historians.

9. The hypothesis of general laws operative in history was carried further in a book which appeared in England twenty years later and exercised an influence in Europe far beyond its intrinsic merit, Buckle's History of Civilisation in England (1857-61). Buckle owed much to Comte, and followed him, or rather outdid him, in regarding intellect as the most important factor conditioning the upward development of man, so that progress, according to him, consisted in the victory of the intellectual over the moral laws.

10. The tendency of Comte and Buckle to assimilate history to the sciences of nature by reducing it to general "laws," derived stimulus and plausibility from the vista offered by the study of statistics, in which the Belgian Quetelet, whose book Sur l'homme appeared in 1835, discerned endless possibilities. The astonishing uniformities which statistical inquiry disclosed led to the belief that it was only a question of collecting a sufficient amount of statistical material, to enable us to predict how a given social group will act in a particular case. Bourdeau, a disciple of this school, looks forward to the time when historical science will become entirely quantitative. The actions of prominent individuals, which are generally considered to have altered or determined the course of things, are obviously not amenable to statistical computation or explicable by general laws. Thinkers like Buckle sought to minimise their importance or explain them away.

11. These indications may suffice to show that the new efforts to interpret history which marked the first half of the nineteenth century were governed by conceptions closely related to those which were current in the field of natural science and which resulted in the doctrine of evolution. The genetic principle, progressive development, general laws, the significance of time, the conception of society as an organic aggregate, the metaphysical theory of history as the self-evolution of spirit,--all these ideas show that historical inquiry had been advancing independently on somewhat parallel lines to the sciences of nature. It was necessary to bring this out in order to appreciate the influence of Darwinism.

12. In the course of the dozen years which elapsed between the

appearances of The Origin of Species (observe that the first volume of Buckle's work was published just two years before) and of The Descent of Man (1871), the hypothesis of Lamarck that man is the co-descendant with other species of some lower extinct form was admitted to have been raised to the rank of an established fact by most thinkers whose brains were not working under the constraint of theological authority.

One important effect of the discovery of this fact (I am not speaking now of the Darwinian explanation) was to assign to history a definite place in the coordinated whole of knowledge, and relate it more closely to other sciences. It had indeed a defined logical place in systems such as Hegel's and Comte's; but Darwinism certified its standing convincingly and without more ado. The prevailing doctrine that man was created ex abrupto had placed history in an isolated position, disconnected with the sciences of nature. Anthropology, which deals with the animal anthropos, now comes into line with zoology, and brings it into relation with history.[239] Man's condition at the present day is the result of a series of transformations, going back to the most primitive phase of society, which is the ideal (unattainable) beginning of history. But that beginning had emerged without any breach of continuity from a development which carries us back to a quadrimane ancestor, still further back (according to Darwin's conjecture) to a marine animal of the ascidian type, and then through remoter periods to the lowest form of organism. It is essential in this theory that though links have been lost there was no break in the gradual development; and this conception of a continuous progress in the evolution of life, resulting in the appearance of uncivilised Anthropos, helped to reinforce, and increase a belief in, the conception of the history of civilised Anthropos as itself also a continuous progressive development.

13. Thus the diffusion of the Darwinian theory of the origin of man, by emphasising the idea of continuity and breaking down the barriers between the human and animal kingdoms, has had an important effect in establishing the position of history among the sciences which deal with telluric development. The perspective of history is merged in a larger perspective of development. As one of the objects of biology is to find the exact steps in the genealogy of man from the lowest organic form, so the scope of history is to determine the stages in the unique causal series from the most rudimentary to the present state of human civilisation.

It is to be observed that the interest in historical research implied by this conception need not be that of Comte. In the Positive Philosophy history is part of sociology; the interest in it is to discover the sociological laws. In the view of which I have just spoken, history is permitted to be an end in itself; the reconstruction of the genetic process is an independent interest. For the purpose of the reconstruction, sociology, as well as physical geography, biology, psychology, is necessary; the sociologist and the historian play into each other's hands; but the object of the former is to establish generalisations; the aim of the latter is to trace in detail a singular causal sequence.

14. The success of the evolutional theory helped to discredit the assumption or at least the invocation of transcendent causes. Philosophically of course it is compatible with theism, but historians have for the most part desisted from invoking the naive conception of a "god in history" to explain historical movements. A historian may be a theist; but, so far as his work is concerned, this particular belief is otiose. Otherwise indeed (as was remarked above) history could not be a science; for with a deus ex machina who can be brought on the stage to solve difficulties scientific treatment is a farce. The transcendent element had appeared in a more subtle form through the influence of German philosophy. I noticed how Ranke is prone to refer to ideas as if they were transcendent existences manifesting themselves in the successive movements of history. It is intelligible to speak of certain ideas as controlling, in a given period,--for instance, the idea of nationality; but from the scientific point of view, such ideas have no existence outside the minds of individuals and are purely psychical forces; and a historical "idea," if it does not exist in this form, is merely a way of expressing a synthesis of the historian himself.

15. From the more general influence of Darwinism on the place of history in the system of human knowledge, we may turn to the influence of the principles and methods by which Darwin explained development. It had been recognised even by ancient writers (such as Aristotle and Polybius) that physical circumstances (geography, climate) were factors conditioning the character and history of a race or society. In the sixteenth century Bodin emphasised these factors, and many subsequent writers took them into account. The investigations of Darwin, which brought them into the

foreground, naturally promoted attempts to discover in them the chief key to the growth of civilisation. Comte had expressly denounced the notion that the biological methods of Lamarck could be applied to social man. Buckle had taken account of natural influences, but had relegated them to a secondary plane, compared with psychological factors. But the Darwinian theory made it tempting to explain the development of civilisation in terms of "adaptation to environment," "struggle for existence," "natural selection," "survival of the fittest," etc.[240]

The operation of these principles cannot be denied. Man is still an animal, subject to zoological as well as mechanical laws. The dark influence of heredity continues to be effective; and psychical development had begun in lower organic forms,--perhaps with life itself. The organic and the social struggles for existence are manifestations of the same principle. Environment and climatic influence must be called in to explain not only the differentiation of the great racial sections of humanity, but also the varieties within these sub-species and, it may be, the assimilation of distinct varieties. Ritter's Anthropogeography has opened a useful line of research. But on the other hand, it is urged that, in explaining the course of history, these principles do not take us very far, and that it is chiefly for the primitive ultra-prehistoric period that they can account for human development. It may be said that, so far as concerns the actions and movements of men which are the subject of recorded history, physical environment has ceased to act mechanically, and in order to affect their actions must affect their wills first; and that this psychical character of the causal relations substantially alters the problem. The development of human societies, it may be argued, derives a completely new character from the dominance of the conscious psychical element, creating as it does new conditions (inventions, social institutions, etc.) which limit and counteract the operation of natural selection, and control and modify the influence of physical environment. Most thinkers agree now that the chief clews to the growth of civilisation must be sought in the psychological sphere. Imitation, for instance, is a principle which is probably more significant for the explanation of human development than natural selection. Darwin himself was conscious that his principles had only a very restricted application in this sphere, as is evident from his cautious and tentative remarks in the 5th chapter of his Descent of Man. He applied natural selection to the growth of the intellectual faculties and of the fundamental social instincts, and also to the differentiation of the great races or "sub-species" (Caucasian, African, etc.)

which differ in anthropological character.[241]

16. But if it is admitted that the governing factors which concern the student of social development are of the psychical order, the preliminary success of natural science in explaining organic evolution by general principles encouraged sociologists to hope that social evolution could be explained on general principles also. The idea of Condorcet, Buckle, and others, that history could be assimilated to the natural sciences was powerfully reinforced, and the notion that the actual historical process, and every social movement involved in it, can be accounted for by sociological generalisations, so-called "laws," is still entertained by many, in one form or another. Dissentients from this view do not deny that the generalisations at which the sociologist arrives by the comparative method, by the analysis of social factors, and by psychological deduction may be an aid to the historian; but they deny that such uniformities are laws or contain an explanation of the phenomena. They can point to the element of chance coincidence. This element must have played a part in the events of organic evolution, but it has probably in a larger measure helped to determine events in social evolution. The collision of two unconnected sequences may be fraught with great results. The sudden death of a leader or a marriage without issue, to take simple cases, has again and again led to permanent political consequences. More emphasis is laid on the decisive actions of individuals, which cannot be reduced under generalisations and which deflect the course of events. If the significance of the individual will had been exaggerated to the neglect of the collective activity of the social aggregate before Condorcet, his doctrine tended to eliminate as unimportant the roles of prominent men, and by means of this elimination it was possible to found sociology. But it may be urged that it is patent on the face of history that its course has constantly been shaped and modified by the wills of individuals,[242] which are by no means always the expression of the collective will; and that the appearance of such personalities at the given moments is not a necessary outcome of the conditions and cannot be deduced. Nor is there any proof that, if such and such an individual had not been born, some one else would have arisen to do what he did. In some cases there is no reason to think that what happened need ever have come to pass. In other cases, it seems evident that the actual change was inevitable, but in default of the man who initiated and guided it, it might have been postponed, and, postponed or not, might have borne a different cachet. I may illustrate by an instance which has just come under my

notice. Modern painting was founded by Giotto, and the Italian expedition of Charles VIII, near the close of the sixteenth century, introduced into France the fashion of imitating Italian painters. But for Giotto and Charles VIII, French painting might have been very different. It may be said that "if Giotto had not appeared, some other great imitator would have played a role analogous to his, and that without Charles VIII there would have been the commerce with Italy, which in the long run would have sufficed to place France in relation with Italian artists. But the equivalent of Giotto might have been deferred for a century and probably would have been different; and commercial relations would have required ages to produce the rayonnement imitatif of Italian art in France, which the expedition of the royal adventurer provoked in a few years."[243] Instances furnished by political history are simply endless. Can we conjecture how events would have moved if the son of Philip of Macedon had been an incompetent? The aggressive action of Prussia which astonished Europe in 1740 determined the subsequent history of Germany; but that action was anything but inevitable; it depended entirely on the personality of Frederick the Great.

Hence it may be argued that the action of individual wills is a determining and disturbing factor, too significant and effective to allow history to be grasped by sociological formulae. The types and general forms of development which the sociologist attempts to disengage can only assist the historian in understanding the actual course of events. It is in the special domains of economic history and Culturgeschichte which have come to the front in modern times that generalisation is most fruitful, but even in these it may be contended that it furnishes only partial explanations.

17. The truth is that Darwinism itself offers the best illustration of the insufficiency of general laws to account for historical development. The part played by coincidence, and the part played by individuals--limited by, and related to, general social conditions--render it impossible to deduce the course of the past history of man or to predict the future. But it is just the same with organic development. Darwin (or any other zoologist) could not deduce the actual course of evolution from general principles. Given an organism and its environment, he could not show that it must evolve into a more complex organism of a definite predetermined type; knowing what it has evolved into, he could attempt to discover and assign the determining causes. General principles do not account for a particular sequence; they

embody necessary conditions; but there is a chapter of accidents too. It is the same in the case of history.

18. Among the evolutional attempts to subsume the course of history under general syntheses, perhaps the most important is that of Lamprecht, whose "kulturhistorische" attempt to discover and assign the determining causes. German history, exhibits the (indirect) influence of the Comtist school. It is based upon psychology, which, in his views, holds among the sciences of mind (Geisteswissenschaften) the same place (that of a Grundwissenschaft) which mechanics holds among the sciences of nature. History, by the same comparison, corresponds to biology, and, according to him, it can only become scientific if it is reduced to general concepts (Begriffe). Historical movements and events are of a psychical character, and Lamprecht conceives a given phase of civilisation as "a collective psychical condition (seelischer Gesamtzustand)" controlling the period, "a diapason which penetrates all psychical phenomena and thereby all historical events of the time."[244] He has worked out a series of such phases, "ages of changing psychical diapason," in his Deutsche Geschichte, with the aim of showing that all the feelings and actions of each age can be explained by the diapason; and has attempted to prove that these diapasons are exhibited in other social developments, and are consequently not singular but typical. He maintains further that these ages succeed each other in a definite order; the principle being that the collective psychical development begins with the homogeneity of all the individual members of a society and, through heightened psychical activity, advances in the form of a continually increasing differentiation of the individuals (this is akin to the Spencerian formula). This process, evolving psychical freedom from psychical constraint, exhibits a series of psychical phenomena which define successive periods of civilisation. The process depends on two simple principles, that no idea can disappear without leaving behind it an effect or influence, and that all psychical life, whether in a person or a society, means change, the acquisition of new mental contents. It follows that the new have to come to terms with the old, and this leads to a synthesis which determines the character of a new age. Hence the ages of civilisation are defined as the "highest concepts for subsuming without exception all psychical phenomena of the development of human societies, that is, of all historical events."[245] Lamprecht deduces the idea of a special historical science, which might be called "historical ethnology," dealing with the ages of civilisation, and bearing the same relation to (descriptive or narrative) history

as ethnology to ethnography. Such a science obviously corresponds to Comte's social dynamics, and the comparative method, on which Comte laid so much emphasis, is the principal instrument of Lamprecht.

19. I have dwelt on the fundamental ideas of Lamprecht, because they are not yet widely known in England, and because his system is the ablest product of the sociological school of historians. It carries the more weight as its author himself is a historical specialist, and his historical syntheses deserve the most careful consideration. But there is much in the process of development which on such assumptions is not explained, especially the initiative of individuals. Historical development does not proceed in a right line, without the choice of diverging. Again and again, several roads are open to it, of which it chooses one--why? On Lamprecht's method, we may be able to assign the conditions which limit the psychical activity of men at a particular stage of evolution, but within those limits the individual has so many options, such a wide room for moving, that the definition of those conditions, the "psychical diapasons," is only part of the explanation of the particular development. The heel of Achilles in all historical speculations of this class has been the role of the individual.

The increasing prominence of economic history has tended to encourage the view that history can be explained in terms of general concepts or types. Marx and his school based their theory of human development on the conditions of production, by which, according to them, all social movements and historical changes are entirely controlled. The leading part which economic factors play in Lamprecht's system is significant, illustrating the fact that economic changes admit most readily this kind of treatment, because they have been less subject to direction or interference by individual pioneers.

Perhaps it may be thought that the conception of social environment (essentially psychical), on which Lamprecht's "psychical diapasons" depend, is the most valuable and fertile conception that the historian owes to the suggestion of the science of biology--the conception of all particular historical actions and movements as (1) related to and conditioned by the social environment, and (2) gradually bringing about a transformation of that environment. But no given transformation can be proved to be necessary (predetermined). And types of development do not represent laws; their meaning and value lie in the help they may give to the historian, in investigating a certain period of civilisation, to enable him to discover the inter-relations among the diverse features which it presents. They are, as some one has said, an instrument of heuretic method.

20. The man engaged in special historical researches--which have been pursued unremittingly for a century past, according to scientific methods of investigating evidence (initiated by Wolf, Niebuhr, Ranke)--have for the most part worked on the assumptions of genetic history or at least followed in the footsteps of those who fully grasped the genetic point of view. But their aim has been to collect and sift evidence, and determine particular facts; comparatively few have given serious thought to the lines of research and the speculations which have been considered in this paper. They have been reasonably shy of compromising their work by applying theories which are still much debated and immature. But historiography cannot permanently evade the questions raised by these theories. One may venture to say that no historical change or transformation will be fully understood until it is explained how social environment acted on the individual components of the society (both immediately and by heredity), and how the individuals reacted upon their environment. The problem is psychical, but it is analogous to the main problem of the biologist.

FOOTNOTES:

[Footnote 238: A society presents suggestive analogies with an organism, but it certainly is not an organism, and sociologists who draw inferences from the assumption of its organic nature must fall into error. A vital organism and a society are radically distinguished by the fact that the individual components of the former, namely the cells, are morphologically as well as functionally differentiated, whereas the individuals which compose a society are morphologically homogeneous and only functionally differentiated. The resemblances and the differences are worked out in E. de Majewski's striking book, La Science de la Civilisation. Paris. 1908.]

[Footnote 239: It is to be observed that history is (not only different in scope but) not co-extensive with anthropology in time. For it deals only with the development of man in societies, whereas anthropology includes in its definition the proto-anthropic period when anthropos was still non-social, whether he lived in herds like the chimpanzee, or alone like the male ourang-outang. (It has been well shown by Majewski that congregations--herds, flocks, packs, &c.--of animals are not societies; the characteristic of a society is differentiation of function. Bee hives, ant hills, may be called quasi-societies; but in their case the classes which perform distinct functions are morphologically different.)]

[Footnote 240: Recently O. Seeck has applied these principles to the decline of Graeco-Roman civilisation in his Untergang der antiken Welt, 2 vols., Berlin, 1895, 1901.]

[Footnote 241: Darwinian formulae may be suggestive by way of analogy. For instance, it is characteristic of social advance that a multitude of inventions, schemes and plans are framed which are never carried out, similar to, or designed for the same end as, an invention or plan which is actually adopted because it has chanced to suit better the particular conditions of the hour (just as the works accomplished by an individual statesman, artist or savant are usually only a residue of the numerous projects conceived by his brain). This process in which so much abortive production occurs is analogous to elimination by natural selection.]

[Footnote 242: We can ignore here the metaphysical question of freewill and determinism. For the character of the individual's brain depends in any case on ante-natal accidents and coincidences, and so it may be said that the role of individuals ultimately depends on chance,--the accidental coincidence of independent sequences.]

[Footnote 243: I have taken this example from G. Tarde's La logique sociale (p. 403), Paris, 1904, where it is used for quite a different purpose.]

[Footnote 244: Die kulturhistorische Methode, Berlin, 1900, p. 26.]

[Footnote 245: Ibid. pp. 28, 29.]

X

DARWINISM AND SOCIOLOGY

How has our conception of social phenomena, and of their history, been affected by Darwin's conception of Nature and the laws of its transformation? To what extent and in what particular respects have the discoveries and hypotheses of the author of The Origin of Species aided the efforts of those who have sought to construct a science of society?

To such a question it is certainly not easy to give any brief or precise answer. We find traces of Darwinism almost everywhere. Sociological systems differing widely from each other have laid claim to its authority; while, on the other hand, its influence has often made itself felt only in combination with other influences. The Darwinian thread is worked into a hundred patterns along with other threads.

To deal with the problem, we must, it seems, first of all distinguish the more general conclusions in regard to the evolution of living beings, which are the outcome of Darwinism, from the particular explanations it offers of the ways and means by which that evolution is effected. That is to say, we must, as far as possible, estimate separately the influence of Darwin as an evolutionist and Darwin as a selectionist.

The nineteenth century, said Cournot, has witnessed a mighty effort. From divers quarters there has been a methodical reaction against the persistent dualism of the Cartesian tradition, which was itself the unconscious heir of the Christian tradition. Even the philosophy of the eighteenth century, materialistic as were for the most part the tendencies of its leaders, seemed to revere man as a being apart, concerning whom laws might be formulated ?priori. To bring him down from his pedestal there was needed the marked predominance of positive researches wherein no account was taken of the "pride of man." There can be no doubt that Darwin has done much to familiarise us with this attitude. Take for instance the first part of The Descent of Man: it is an accumulation of typical facts, all tending to diminish the distance between us and our brothers, the lower animals. One might say that the naturalist had here taken as his motto, "Whosoever shall exalt himself shall be abased; and he that shall humble himself shall be exalted." Homologous structures, the survival in man of certain organs of animals, the rudiments in the animal of certain human faculties, a multitude of facts of this sort, led Darwin to the conclusion that there is no ground for supposing that the "king of the universe" is exempt from universal laws. Thus belief in the imperium in imperio has been, as it were, whittled away by the progress of the naturalistic spirit, itself continually strengthened by the conquests of the natural sciences. The tendency may, indeed, drag the social sciences into overstrained analogies, such, for instance, as the assimilation of societies to organisms. But it will, at least, have had the merit of helping sociology to shake off the pre-conception that the groups formed by men are artificial, and that history is completely at the mercy of chance. Some years before the appearance of The Origin of Species, August Comte had pointed out the importance, as regards the unification of positive knowledge, of the conviction that the social world, the last refuge of spiritualism, is itself subject to determinism. It cannot be doubted that the movement of thought which Darwin's discoveries promoted contributed to the spread of this conviction, by breaking down the traditional barrier which cut man off from Nature.

But Nature, according to modern naturalists, is no immutable thing: it is rather perpetual movement, continual progression. Their discoveries batter a breach directly into the Aristotelian notion of species; they refuse to see in the animal world a collection of immutable types, distinct from all eternity, and corresponding, as Cuvier said, to so many particular thoughts of the Creator. Darwin especially congratulated himself upon having been able to deal this doctrine the coup de grate: immutability is, he says, his chief enemy; and he is concerned to show--therein following up Lyell's work--that everything in the organic world, as in the inorganic, is explained by insensible but incessant transformations. "Nature makes no leaps"--"Nature knows no gaps": these two dicta form, as it were, the two landmarks between which Darwin's idea of transformation is worked out. That is to say, the development of Darwinism is calculated to further the application of the philosophy of Becoming to the study of human institutions.

The progress of the natural sciences thus brings unexpected reinforcements to the revolution which the progress of historical discipline had begun. The first attempt to constitute an actual science of social phenomena--that, namely, of the economists--had resulted in laws which were called natural, and which were believed to be eternal and universal, valid for all times and all places. But this perpetuality, brother, as Knies said, of the immutability of the old zoology, did not long hold out against the ever-swelling tide of the historical movement. Knowledge of the transformations that had taken place in language, of the early phases of the family, of religion, of property, had all favoured the revival of the Heraclitean view: [Greek: panta rei]. As to the categories of political economy, it was soon to be recognised, as by Lasalle, that they too are only historical. The philosophy of history, moreover, gave expression under various forms to the same tendency. Hegel declares that "all that is real is rational," but at the same time he shows that all that is real is ephemeral, and that for history there is nothing fixed beneath the sun. It is this sense of universal evolution that Darwin came with fresh authority to enlarge. It was in the name of biological facts themselves that he taught us to see only slow metamorphoses in the history of institutions, and to be always on the outlook for survivals side by side with rudimentary forms. Anyone who reads Primitive Culture, by Tylor,--a writer closely connected with Darwin-- will be able to estimate the services which these cardinal ideas were to render to the social sciences when the age of comparative research had succeeded to that of ?priori construction.

Let us note, moreover, that the philosophy of Becoming in passing through the Darwinian biology became, as it were, filtered; it got rid of those traces of finalism, which, under different forms, it had preserved through all the systems of German Romanticism. Even in Herbert Spencer, it has been plausibly argued, one can detect something of that sort of mystic confidence in forces spontaneously directing life, which forms the very essence of those systems. But Darwin's observations were precisely calculated to render such an hypothesis futile. At first people may have failed to see this; and we call to mind the ponderous sarcasms of Flourens when he objected to the theory of Natural Selection that it attributed to nature a power of free choice. "Nature endowed with will! That was the final error of last century; but the nineteenth no longer deals in personifications."[246] In fact Darwin himself put his readers on their guard against the metaphors he was obliged to use. The processes by which he explains the survival of the fittest are far from affording any indication of the design of some transcendent breeder. Nor, if we look closely, do they even imply immanent effort in the animal; the sorting out can be brought about mechanically, simply by the action of the environment. In this connection Huxley could with good reason maintain that Darwin's originality consisted in showing how harmonies which hitherto had been taken to imply the agency of intelligence and will could be explained without any such intervention. So, when later on, objective sociology declares that, even when social phenomena are in question, all finalist preconceptions must be distrusted if a science is to be constituted, it is to Darwin that its thanks are due; he had long been clearing paths for it which lay well away from the old familiar road trodden by so many theories of evolution.

This anti-finalist doctrine, when fully worked out, was, moreover, calculated to aid in the needful dissociation of two notions: that of evolution and that of progress. In application to society these had long been confounded; and, as a consequence, the general idea seemed to be that only one type of evolution was here possible. Do we not detect such a view in Comte's sociology, and perhaps even in Herbert Spencer's? Whoever, indeed, assumes an end for evolution is naturally inclined to think that only one road leads to that end. But those whose minds the Darwinian theory has enlightened are aware that the transformations of living beings depend primarily upon their conditions, and that it is these conditions which are the agents of selection from among individual variations. Hence, it immediately follows that transformations are not necessarily improvements. Here, Darwin's thought hesitated. Logically his theory proves, as Ray Lankester pointed out, that the struggle for existence may have as its outcome degeneration as well as amelioration: evolution may be regressive as well as progressive. Then, too--and this is especially to be borne in mind--each species takes its good where it finds it, seeks its own path and survives as best it can. Apply this notion to society and you arrive at the theory of multilinear evolution. Divergencies will no longer surprise you. You will be forewarned not to apply to all civilisations the same measure of progress, and you will recognise that types of evolution may differ just as social species themselves differ. Have we not here one of the conceptions which mark off sociology proper from the old philosophy of history?

* * * * *

But if we are to estimate the influence of Darwinism upon sociological conceptions, we must not dwell only upon the way in which Darwin impressed the general notion of evolution upon the minds of thinkers. We must go into details. We must consider the influence of the particular theories by which he explained the mechanism of this evolution. The name of the author of The Origin of Species has been especially attached, as everyone knows, to the doctrines of "natural selection" and of "struggle for existence," completed by the notion of "individual variation." These doctrines were turned to account by very different schools of social philosophy. Pessimistic and optimistic, aristocratic and democratic, individualistic and socialistic systems were to war with each other for years by casting scraps of Darwinism at each other's heads.

It was the spectacle of human contrivance that suggested to Darwin his conception of natural selection. It was in studying the methods of pigeon breeders that he divined the processes by which nature, in the absence of design, obtains analogous results in the differentiation of types. As soon as the importance of artificial selection in the transformation of species of animals was understood, reflection naturally turned to the human species, and the question arose, How far do men observe, in connection with themselves, those laws of which they make practical application in the case of animals? Here we come upon one of the ideas which guided the researches of Gallon, Darwin's cousin. The author of Inquiries into Human Faculty and its Development,[247] has often expressed his surprise that, considering all the precautions taken, for example, in the breeding of horses, none whatever are taken in the breeding of the human species. It seems to be forgotten that the species suffers when the "fittest" are not able to perpetuate their type. Ritchie, in his Darwinism and Politics[248] reminds us of Darwin's remark that the institution of the peerage might be defended on the ground that peers, owing to the prestige they enjoy, are enabled to select as wives "the most beautiful and charming women out of the lower ranks."[249] But, says Galton, it is as often as not "heiresses" that they pick out, and birth statistics seem to show that these are either less robust or less fecund than others. The truth is that considerations continue to preside over marriage which are entirely foreign to the improvement of type, much as this is a condition of general progress. Hence the importance of completing Odin's and De Candolle's statistics which are designed to show how characters are incorporated in organisms, how they are transmitted, how lost, and according to what law eugenic, elements depart from the mean or return to it.

But thinkers do not always content themselves with undertaking merely the minute researches which the idea of Selection suggests. They are eager to defend this or that thesis. In the name of this idea certain social anthropologists have recast the conception of the process of civilisation, and have affirmed that Social Selection generally works against the trend of Natural Selection. Vacher de Lapouge--following up an observation by Broca on the point--enumerates the various institutions, or customs, such as the celibacy of priests and military conscription, which cause elimination or sterilisation of the bearers of certain superior qualities, intellectual or physical. In a more general way he attacks the democratic movement, a movement, as P. Bourget says, which is "anti-physical" and contrary to the natural laws of progress; though it has been inspired "by the dreams of that most visionary of all centuries, the eighteenth."[250] The "Equality" which levels down and mixes (justly condemned, he holds, by the Comte de Gobineau), prevents the aristocracy of the blond dolichocephales from holding the position and playing the part which, in the interests of all, should belong to them. Otto Ammon, in his Natural Selection in Man, and in The Social Order and its Natural Bases,[251] defended analogous doctrines in Germany; setting the curve representing frequency of talent over against that of income, he attempted to show that all democratic measures which aim at promoting the rise in the social scale of the talented are useless, if not dangerous; that they only increase the panmixia, to the great detriment of the species and of society.

Among the aristocratic theories which Darwinism has thus inspired we must reckon that of Nietzsche. It is well known that in order to complete his philosophy he added biological studies to his philological; and more than once in his remarks upon the Wille zur Macht he definitely alludes to Darwin; though it must be confessed that it is generally in order to proclaim the insufficiency of the processes by which Darwin seeks to explain the genesis of species. Nevertheless, Nietzsche's mind is completely possessed by an ideal of Selection. He, too, has a horror of panmixia. The naturalists' conception of "the fittest" is joined by him to that of the "hero" of romance to furnish a basis for his doctrine of the Superman. Let us hasten to add, moreover, that at the very moment when support was being sought in the theory of Selection for the various forms of the aristocratic doctrine, those same forms were being battered down on another side by means of that very theory. Attention was drawn to the fact that by virtue of the laws which Darwin himself had discovered isolation leads to etiolation. There is a risk that the privilege which withdraws the privileged elements of Society from competition will cause them to degenerate. In fact, Jacoby in his Studies in Selection, in connexion with Heredity in Man,[252] concludes that "sterility, mental debility, premature death and, finally, the extinction of the stock were not specially and exclusively the fate of sovereign dynasties; all privileged classes, all families in exclusively elevated positions share the fate of reigning families, although in a minor degree and in direct proportion to the loftiness of their social standing. From the mass of human beings spring individuals, families, races, which tend to raise themselves above the common level; painfully they climb the rugged heights, attain the summits of power, of wealth, of intelligence, of talent, and then, no sooner are they there than they topple down and disappear in gulfs of mental and physical degeneracy." The demographical researches of Hansen[253] (following up and completing Dumont's) tended, indeed, to show that urban as well as feudal aristocracies, burgher classes as well as noble castes, were liable to become effete. Hence it might well be concluded that the democratic movement, operating as it does to break down class barriers, was promoting instead of impeding human selection.

* * * * *

So we see that, according to the point of view, very different conclusions have been drawn from the application of the Darwinian idea of Selection to human society. Darwin's other central idea, closely bound up with this, that, namely, of the "struggle for existence" also has been diversely utilised. But discussion has chiefly centered upon its signification. And while some endeavour to extend its application to everything, we find others trying to limit its range. The conception of a "struggle for existence" has in the present day been taken up into the social sciences from natural science, and adopted. But originally it descended from social science to natural. Darwin's law is, as he himself said, only Malthus' law generalised and extended to the animal world: a growing disproportion between the supply of food and the number of the living is the fatal order whence arises the necessity of universal struggle, a struggle which, to the great advantage of the species, allows only the best equipped individuals to survive. Nature is regarded by Huxley as an immense arena where all living beings are gladiators.[254]

Such a generalisation was well adapted to feed the stream of pessimistic thought; and it furnished to the apologists of war, in particular, new arguments, weighted with all the authority which in these days attaches to scientific deliverances. If people no longer say, as Bonald did, and Moltke after him, that war is a providential fact, they yet lay stress on the point that it is a natural fact. To the peace party Dragomirov's objection is urged that its attempts are contrary to the fundamental laws of nature, and that no sea wall can hold against breakers that come with such gathered force.

But in yet another quarter Darwinism was represented as opposed to philanthropic intervention. The defenders of the orthodox political economy found in it support for their tenets. Since in the organic world universal struggle is the condition of progress, it seemed obvious that free competition must be allowed to reign unchecked in the economic world. Attempts to curb it were in the highest degree imprudent. The spirit of Liberalism here seemed in conformity with the trend of nature: in this respect, at least, contemporary naturalism, offspring of the discoveries of the nineteenth century, brought reinforcements to the individualist doctrine, begotten of the speculations of the eighteenth: but only, it appeared, to turn mankind away for ever from humanitarian dreams. Would those whom such conclusions repelled be content to oppose to nature's imperatives only the protests of the heart? There were some who declared, like Brunetie, that the laws in question, valid though they might be for the animal kingdom, were not applicable to the human. And so a return was made to the classic dualism. This indeed seems to be the line that Huxley took, when, for instance, he opposed to the cosmic process an ethical process which was its reverse.

But the number of thinkers whom this antithesis does not satisfy grows daily. Although the pessimism which claims authorisation from Darwin's doctrines is repugnant to them, they still are unable to accept the dualism which leaves a gulf between man and nature. And their endeavour is to link the two by showing that while Darwin's laws obtain in both kingdoms, the conditions of their application are not the same: their forms, and, consequently, their results, vary with the varying mediums in which the struggle of living beings takes place, with the means these beings have at disposal, with the ends even which they propose to themselves.

Here we have the explanation of the fact that among determined opponents of war partisans of the "struggle for existence" can be found: there are disciples of Darwin in the peace party. Novicow, for example, admits the "combat universel" of which Le Dantec[255] speaks; but he remarks that at different stages of evolution, at different stages of life the same weapons are not necessarily employed. Struggles of brute force, armed hand to hand conflicts, may have been a necessity in the early phases of human societies. Nowadays, although competition may remain inevitable and indispensable, it can assume milder forms. Economic rivalries, struggles between intellectual influences, suffice to stimulate progress: the processes which these admit are, in the actual state of civilisation, the only ones which attain their end without waste, the only ones logical. From one end to the other of the ladder of life, struggle is the order of the day; but more and more as the higher rungs are reached, it takes on characters which are proportionately more "humane."

Reflections of this kind permit the introduction into the economic order of limitations to the doctrine of "laisser faire, laisser passer." This appeals, it is said, to the example of nature where creatures, left to themselves, struggle without truce and without mercy; but the fact is forgotten that upon industrial battlefields the conditions are different. The competitors here are not left simply to their natural energies: they are variously handicapped. A rich store of artificial resources exists in which some participate and others do not. The sides then are unequal; and as a consequence the result of the struggle is falsified. "In the animal world," said De Laveleye,[256] criticising Spencer, "the fate of each creature is determined by its individual qualities; whereas in civilised societies a man may obtain the highest position and the most beautiful wife because he is rich and well-born, although he may be ugly, idle or improvident; and then it is he who will perpetuate the species. The wealthy man, ill constituted, incapable, sickly, enjoys his riches and establishes his stock under the protection of the laws." Haycraft in England and Jentsch in Germany have strongly emphasised these "anomalies," which nevertheless are the rule. That is to say that even from a Darwinian point of view all social reforms can readily be justified which aim at diminishing, as Wallace said, inequalities at the start.

But we can go further still. Whence comes the idea that all measures inspired by the sentiment of solidarity are contrary to Nature's trend? Observe her carefully, and she will not give lessons only in individualism. Side by side with the struggle for existence do we not find in operation what Lanessan calls "association for existence." Long ago, Espinas had drawn attention to "societies of animals," temporary or permanent, and to the kind of morality that arose in them. Since then, naturalists have often insisted upon the importance of various forms of symbiosis. Kropotkin in Mutual Aid has chosen to enumerate many examples of altruism furnished by animals to mankind. Geddes and Thomson went so far as to maintain that "Each of the greater steps of progress is in fact associated with an increased measure of subordination of individual competition to reproductive or social ends, and of interspecific competition to co-operative, association."[257] Experience shows, according to Geddes, that the types which are fittest to surmount great obstacles are not so much those who engage in the fiercest competitive struggle for existence, as those who contrive to temper it. From all these observations there resulted, along with a limitation of Darwinian pessimism, some encouragement for the aspirations of the collectivists.

And Darwin himself would, doubtless, have subscribed to these rectifications. He never insisted, like his rival, Wallace, upon the necessity of the solitary struggle of creatures in a state of nature, each for himself and against all. On the contrary, in The Descent of Man, he pointed out the serviceableness of the social instincts, and corroborated Bagehot's statements when the latter, applying laws of physics to politics, showed the great advantage societies derived from intercourse and communion. Again, the theory of sexual evolution which makes the evolution of types depend increasingly upon preferences, judgments, mental factors, surely offers something to qualify what seems hard and brutal in the theory of natural selection.

But, as often happens with disciples, the Darwinians had out-Darwined Darwin. The extravagances of social Darwinism provoked a useful reaction; and thus people were led to seek, even in the animal kingdom, for facts of solidarity which would serve to justify humane effort.

* * * * *

On quite another line, however, an attempt has been made to connect socialist tendencies with Darwinian principles. Marx and Darwin have been confronted; and writers have undertaken to show that the work of the German philosopher fell readily into line with that of the English naturalist and was a development of it. Such has been the endeavour of Ferri in Italy and of Woltmann in Germany, not to mention others. The founders of "scientific socialism" had, moreover, themselves thought of this reconciliation. They make more than one allusion to Darwin in works which appeared after 1859. And sometimes they use his theory to define by contrast their own ideal. They remark that the capitalist system, by giving free course to individual competition, ends indeed in a bellum omnium contra omnes; and they make it clear that Darwinism, thus understood, is as repugnant to them as to Daring.

But it is at the scientific and not at the moral point of view that they place themselves when they connect their economic history with Darwin's work. Thanks to this unifying hypothesis, they claim to have constructed--as Marx does in his preface to Das Kapital--a veritable natural history of social evolution. Engels speaks in praise of his friend Marx as having discovered the true mainspring of history hidden under the veil of idealism and sentimentalism, and as having proclaimed in the primum vivere the inevitableness of the struggle for existence. Marx himself, in Das Kapital, indicated another analogy when he dwelt upon the importance of a general technology for the explanation of this psychology:--a history of tools which would be to social organs what Darwinism is to the organs of animal species. And the very importance they attach to tools, to apparatus, to machines, abundantly proves that neither Marx nor Engels were likely to forget the special characters which mark off the human world from the animal. The former always remains to a great extent an artificial world. Inventions change the face of its institutions. New modes of production revolutionise not only modes of government, but modes even of collective thought. Therefore it is that the evolution of society is controlled by laws special to it, of which the spectacle of nature offers no suggestion.

If, however, even in this special sphere, it can still be urged that the evolution of the material conditions of society is in accord with Darwin's theory, it is because the influence of the methods of production is itself to be explained by the incessant strife of the various classes with each other. So that in the end Marx, like Darwin, finds the source of all progress is in struggle. Both are grandsons of Heraclitus. It sometimes happens, in these days, that the doctrine of revolutionary socialism is contrasted as rude and healthy with what may seem to be the enervating tendency of "solidarist" philanthropy: the apologists of the doctrine then pride themselves above all upon their faithfulness to Darwinian principles.

* * * * *

So far we have been mainly concerned to show the use that social philosophies have made of the Darwinian laws for practical purposes: in order to orientate society towards their ideals each school tries to show that the authority of natural science is on its side. But even in the most objective of theories, those which systematically make abstraction of all political tendencies in order to study the social reality in itself, traces of Darwinism are readily to be found.

Let us take for example Durkheim's theory of Division of Labour.[258] The conclusions he derives from it are that whenever professional specialisation causes multiplication of distinct branches of activity, we get organic solidarity--implying differences--substituted for mechanical solidarity, based upon likenesses. The umbilical cord, as Marx said, which connects the individual consciousness with the collective consciousness is cut. The personality becomes more and more emancipated. But on what does this phenomenon, so big with consequences, itself depend? The author goes to social morphology for the answer: it is, he says, the growing density of population which brings with it this increasing differentiation of activities. But, again, why? Because the greater density, in thrusting men up against each other, augments the intensity of their competition for the means of existence; and for the problems which society thus has to face differentiation of functions presents itself as the gentlest solution.

Here one sees that the writer borrows directly from Darwin. Competition is at its maximum between similars, Darwin had declared; different species, not laying claim to the same food, could more easily coexist. Here lay the explanation of the fact that upon the same oak hundreds of different insects might be found. Other things being equal, the same applies to society. He who finds some unadopted specialty possesses a means of his own for getting a living. It is by this division of their manifold tasks that men contrive not to crush each other. Here we obviously have a Darwinian law serving as intermediary in the explanation of that progress of division of labour which itself explains so much in the social evolution.

And we might take another example, at the other end of the series of sociological systems. G. Tarde is a sociologist with the most pronounced anti-naturalistic views. He has attempted to show that all application of the laws of natural science to society is misleading. In his Opposition Universelle he has directly combatted all forms of sociological Darwinism. According to him the idea that the evolution of society can be traced on the same plan as the evolution of species is chimerical. Social evolution is at the mercy of all kinds of inventions, which by virtue of the laws of imitation modify, through individual to individual, through neighbourhood to neighbourhood, the general state of those beliefs and desires which are the only "quantities" whose variation matters to the sociologist. But, it may be rejoined, that however psychical the forces may be, they are none the less subject to Darwinian laws. They compete with each other; they struggle for the mastery of minds. Between types of ideas, as between organic forms, selection operates. And though it may be that these types are ushered into the arena by unexpected discoveries, we yet recognise in the psychological accidents, which Tarde places at the base of everything, near relatives of those small accidental variations upon which Darwin builds. Thus, accepting Tarde's own representations, it is quite possible to express in Darwinian terms, with the necessary transpositions, one of the most idealistic sociologies that have ever been constructed.

These few examples suffice. They enable us to estimate the extent of the field of influence of Darwinism. It affects sociology not only through the agency of its advocates but through that of its opponents. The questions to which it has given rise have proved no less fruitful than the solutions it has suggested. In short, few doctrines, in the history of social philosophy, will have produced on their passage a finer crop of ideas.

###